LIBRARY
BUENA VISTA UNIVERSITY
610 WEST FOURTH STREET
STORM LAKE, IA 50588

ND 1488 .K82 2005
Kuehni, Rolf G.
Color

Color

Color

An Introduction to Practice and Principles

Second Edition

Rolf G. Kuehni

A JOHN WILEY & SONS, INC., PUBLICATION

Copyright © 2005 by John Wiley & Sons, Inc. All rights reserved.

Published by John Wiley & Sons, Inc., Hoboken, New Jersey.
Published simultaneously in Canada.

No part of this publication may be reproduced, stored in a retrieval system, or transmitted in any form or by any means, electronic, mechanical, photocopying, recording, scanning, or otherwise, except as permitted under Section 107 or 108 of the 1976 United States Copyright Act, without either the prior written permission of the Publisher, or authorization through payment of the appropriate per-copy fee to the Copyright Clearance Center, Inc., 222 Rosewood Drive, Danvers, MA 01923, 978-750-8400, fax 978-646-8600, or on the web at www.copyright.com. Requests to the Publisher for permission should be addressed to the Permissions Department, John Wiley & Sons, Inc., 111 River Street, Hoboken, NJ 07030, (201) 748-6011, fax (201) 748-6008.

Limit of Liability/Disclaimer of Warranty: While the publisher and author have used their best efforts in preparing this book, they make no representations or warranties with respect to the accuracy or completeness of the contents of this book and specifically disclaim any implied warranties of merchantability or fitness for a particular purpose. No warranty may be created or extended by sales representatives or written sales materials. The advice and strategies contained herein may not be suitable for your situation. You should consult with a professional where appropriate. Neither the publisher nor author shall be liable for any loss of profit or any other commercial damages, including but not limited to special, incidental, consequential, or other damages.

For general information on our other products and services please contact our Customer Care Department within the U.S. at 877-762-2974, outside the U.S. at 317-572-3993 or fax 317-572-4002.

Wiley also publishes its books in a variety of electronic formats. Some content that appears in print, however, may not be available in electronic format.

Library of Congress Cataloging-in-Publication Data:

Kuehni, Rolf G.
Color : an introduction to practice and principles / Rolf G. Kuehni. – 2nd ed.
p. cm.
"Wiley-Interscience."
Includes bibliographical references and index.
ISBN 0-471-66006-X (cloth)
1. Color. 2. Color in art. I. Title.
ND1488.K82 2004
535.6–dc22

2004006024

Printed in the United States of America.

10 9 8 7 6 5 4 3 2 1

By convention there is color, by convention sweetness and bitterness, but in reality there are atoms and space.

—Democritus (c. 460 B.C.–c. 370 B.C.) Fragment 125

In his younger years the sixth Ch'an patriarch Hui-neng visited the Fa-hsing temple. He overheard a group of visitors arguing about a banner flapping in the wind. One declared: "The banner is moving." Another insisted: "No, it is the wind that is moving." Hui-neng could not contain himself and interrupted them: "You are both wrong. It is your mind that moves."

—Tun-huang manuscript, Tenth century

Contents

Preface to the Second Edition

This is the third version of an introductory text on the subject of color and color technology. It follows the outline of the first edition of the current book closely, but approximately three quarters of the text has been rewritten for two main reasons: to bring it, in a general manner, up-to-date and to broaden its nontechnical aspects. More stress has been placed on the widening chasm of views about the nature of color: Is it located in nature and physically easily definable or a complex construct of the brain/mind?

Color is a much more encompassing subject than is usually conveyed in standard textbooks on color science and technology. It is part of the very complex vision process whose functioning, despite many advances, remains unknown in detail. There is also the continuing discrepancy between what is known about the physiological processes of color vision and the final results in our conscious experiences. At the same time technological treatment of color is becoming more and more mathematical model driven in a time of economic world competition and of the need to speed up all processes.

The intent remains to provide a relatively simple but technically correct and up-to-date introduction to many aspects of color. The book is intended to be a largely nontechnical text that is reasonably comprehensive, short, and nonmathematical.

Artists, designers, craftsmen, philosophers, psychologists, color technologists, students in many fields with interests in color, or any other person interested in this subject will find first-level answers to many questions related to color as well as insight into the historical development of our knowledge and thinking on the subject. Using the notes, the book can be a stepping stone to more in-depth studies.

Over the years I have become indebted to many people who helped to widen my horizons of this deeply fascinating subject, for which I am grateful. In turn, I hope this book helps open the horizons of many of its readers.

1

Sources of Color

For the normally sighted color is everywhere. In the interior of a dwelling are natural and stained woods, wallpapers, upholstery fabrics, pottery, paintings, plants and flowers, a color television set, and many more things. Out of doors, and depending on the time of the year, there is a riot of colors such as those on an alpine summer meadow, or a sparseness, with olives, browns, garnets, and grays. Colors can be pleasantly subdued, enhancing relaxation, or loud and calling to us from advertising billboards or magazines. Color entices us to eat, consume, or at least to buy.

Color likely has helped us to survive as a species. Our (known) contacts with the world and the universe are by way of our five senses. Persons with a normally functioning visual system obtain what is probably the largest amount of information about the world surrounding them from vision, and color is an important outcome of this flow of information. In the past several thousand years color has blossomed into much more than just a survival and communications tool. We have learned to derive esthetic pleasure from it by way of crafts, design, and art.

The question of the nature of color experiences has puzzled humans since antiquity and has resulted in many and varied answers. The number of different color phenomena in the natural world, from colored sunsets and rainbows and the color of a rose to those of an opal and the glow of phosphors, has made understanding the phenomenon of color rather difficult. The popular view is shaped strongly by our everyday experiences. Bananas are yellow, a ruby-throated hummingbird has a dazzling red patch below his beak, clear water and the sky are blue, and so on. A fabric is dyed with red dye; when painting we use variously colored pigments or we draw with variously colored crayons or ink pens. The rainbow has four colors, or is it

Color: *An Introduction to Practice and Principles, Second Edition,* by Rolf G. Kuehni
ISBN 0471-66006-X Copyright © 2005 John Wiley & Sons, Inc.

six or seven? In a mirror we see colors of objects appearing slightly duller and deeper than in the original. On a winter day, toward evening, shadows look deeply blue. We are told that color illustrations in an art book are printed just with four pigments and that all colors on a TV screen are "made" from a red, a blue, and a green phosphor.

To cope with these confusingly varied sources of color we just disregard them in our everyday languages. An apple is red, the traffic light is red, the rose as seen reflected in a mirror is red, the bar in the bar graph on the computer video display is red, the paint on the brush is red. All of these varied experiences have something in common: redness. We simply attach the perceived phenomenon to the object without bothering about the source or thinking about the nature of color.

We normally experience color as a result of the interaction between light, materials, and our visual apparatus, eye and brain. However, there are also means of having color experiences in the dark, with eyes closed:

Under the influence of migraine headaches
Under the influence of certain drugs
By direct electrical stimulation of certain cells in the brain
By pressing against the eyeballs or hitting the temples moderately hard
By dreaming

In some manner these situations or actions trigger responses in our visual system that have the same result as conventional color stimuli. Such phenomena are not unlike an electronic burglar alarm triggered by on overflying aircraft rather than by a burglar.

The fact that a variety of color stimuli may result in identical experiences for human observers points to color being a subjective phenomenon. Not all philosophers agree; some claim color resides in objects. A convincing theory of what it is in objects such as those mentioned earlier and many others that results in a specific experience of reddishness has not been forthcoming, however. To the author it seems more sensible to assume that colors are not real and the world in front of us is not colored. Newton already made the point about prismatic light by saying "the rays to speak properly are not coloured." On the other hand, this fact seems to fly in the face of our apparent everyday experiences with colorants, materials expressly used to impart color on other materials. But colorants are simply materials that modify reflectance properties.

Color is the result of the activity of one of our five senses, vision. So far, we have not succeeded in defining the essence of the results of sensory activities, emotions, or feelings: what is sweet, what is happy, or blue? Dictionary definitions of color are, therefore, of necessity vague: "a phenomenon of light (as red, brown, pink, or gray) or visual perception that enables one to differentiate otherwise identical objects" (1).

Scientists are equally helpless and have resorted to a circular definition: "perceived color is the attribute of visual perception that can be described by color names: white, gray, black, yellow, orange, brown, red, green, blue, purple, and so on or by combinations of such names" (2). Before considering the difficult subject of the nature of color further it is useful to gain a fuller understanding of the causes of color.

One of the most impressive displays of color occurs when in an otherwise dark room a narrow circular beam of sunlight passes through a glass prism. What leaves the prism is the same light entering it, but on leaving the prism the circular beam has been transformed into a band of light that, when reflected from a white surface, produces in the observer's vision system a multitude of color experiences: the colors known as those of the rainbow. A less elaborate method for viewing these colors is by looking at a compact disk at different angles in the light of a lamp.

A considerable number of processes and materials can result in color experiences. Many have been discovered by artists and craftspersons over the course of millennia, but until recently the underlying causes remained mostly hidden. Colored materials (many used as colorants) are commonly thought to interact in similar ways with light, but their apparent color is in fact caused by a variety of specific phenomena. Nassau has identified and described a total of fifteen causes of color, four dealing with geometrical and physical optics, those remaining with various effects involving electrons in atoms or molecules of materials and causing absorption or emission of light at selected wavebands (3). With the exceptions listed earlier, color phenomena have one common factor: light. Aristotle wrote that the potential of color in materials is activated by light. Goethe called colors "the actions and sufferings of light." The most common source of light is the process of incandescence. Our first step is to gain understanding of the nature of light and incandescence.

LIGHT

Light is a certain kind of electromagnetic radiation, which is a convenient name for the as yet not fully explained phenomenon of energy transport through space. Electromagnetic radiation, depending on its energy content, has different names: X rays capable of passing through our bodies and on prolonged exposure causing serious harm, ultraviolet radiation that can tan or burn our skin, light that we employ to gain visual information about the world around us, infrared radiation that we experience on our skin as warmth or heat, information transmission waves for radio and television, or electricity transmitted and used as a convenient source of energy (Fig. 1.1). Electromagnetic radiation travels at high speed (the speed of light, about 300,000 kilometers per second [km/s]). The eye, our visual sensory organ, is sensitive to a narrow band of electromagnetic radiation, the visible spectrum.

The basic nature of electromagnetic radiation and its mode of transport are not yet fully known. Some experiments show that it travels in the form of waves (comparable to those created when throwing a pebble into a calm pond) or in the form of individual packets of energy, called quanta (singular quantum) or photons. When regarded as waves, the energy content of radiation is usually expressed in terms of wavelength: the shorter the distance between neighboring peaks of waves, the higher the energy content. Wavelength is commonly measured in metric units and the wavelength of visible light ranges from approximately 400 nanometers (nm, billionth of a meter) to 700 nm. When considered as quanta, the energy content is usually expressed as electron volts (eV). Visible electromagnetic radiation can exist

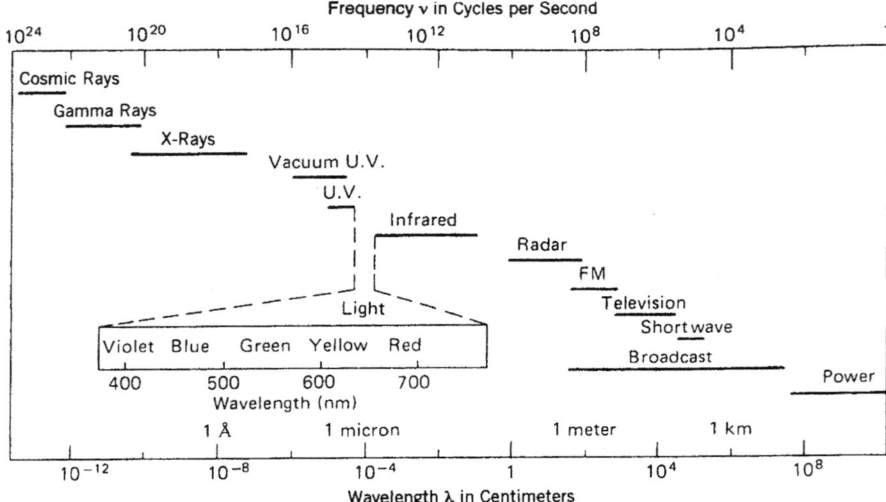

FIGURE 1.1 *Schematic representation of the electromagnetic spectrum.* IES Lighting Handbook, *New York: Illuminating Engineering Society, 1972. Used with permission.*

at a single wavelength (monochromatic) or be a mixture of many wavelengths (polychromatic).

Electromagnetic radiation can interact with matter in different ways:

- *Absorption:* Quanta are absorbed by matter, interact with it in certain ways, and after loss of some energy are reemitted
- *Transmission:* Quanta pass through matter unchanged; certain forms of matter impede the speed of the quanta, which at interfaces of two different kinds of transmitting matter, can result in a change of direction (refraction)
- *Scattering:* Certain matter is impenetrable to quanta and they are scattered or reflected by it, changing direction in the process .
- *Interference:* Quanta can interact with neighboring quanta in certain conditions.

Light is normally produced by a glowing body in a process called incandescence: the sun, for example, a burning wax candle, or an electrically heated tungsten metal coil in a light bulb.

INCANDESCENCE

Incandescence is the shedding of electromagnetic radiation by a very hot material, resulting in light that can give rise to color experiences. Our dominant example of an incandescent body is the sun. The nature of incandescence is most easily observed in the work of a blacksmith (alas, with fewer and fewer opportunities to do so). An iron rod or a horseshoe, placed in an intense coal fire, will as it heats up begin to give off a dull reddish glow. When viewing it in the dark, we recognize it as the source of reddish light. As the temperature of the metal increases so does the intensity of the emitted light. Simultaneously, reddishness diminishes and the object becomes "white-hot." With a further increase in temperature, it eventually assumes a bluish white appearance. Energy is absorbed by the horseshoe from the fire and emitted in visible form by the glowing metal. The imparted energy can have many sources: thermonuclear in the case of the sun; electrical in the case of a light bulb, chemical in the case of burning coal. All elements can, in proper conditions, be made to show incandescence, as can many inorganic molecules. Organic molecules (those containing carbon), are usually destroyed before they show incandescence, with incandescence produced by their decomposition products (say, in the case of candle wax). The nature of the emitted energy depends on the form of the incandescent material: gaseous substances emit energy in one or more distinct bands; incandescent liquids and solids emit energy across broad spectrum bands.

What is the explanation for energy absorption and incandescence? The accepted theory is based on an atomic model of matter, with protons and neutrons in the central nucleus, and electrons located in shells around the nucleus. Each of the shells has limited spaces for electrons. Shells that are filled to their limit or where electrons are in pairs are in a relatively stable state. As the atom or molecule absorbs energy, it passes through various stages of excitation. Each stage involves the electron(s) of the outermost shell. Absorption of energy will raise the excitable outer electron(s) to the next rung on an excitation ladder. At any given time the assembly of atoms or molecules in matter is not only absorbing energy but also shedding it: while in some atoms or molecules the outer electrons are being raised to the next level of excitation, in others they fall back one or more rungs to bring the atom into equilibrium with the average energy content of the surrounding matter. As mentioned, the shed energy is in the form of quanta or waves. If the shed energy is such that its wavelength falls between 400 and 700 nm, we sense it as light. At other levels they fall into other areas in the electromagnetic spectrum, such as ultraviolet or infrared.

The energy rungs possible derive from strict physical laws, and there are many rungs on the energy ladder of an atom or molecule. Electrons can cascade back in a variety of the ways, but there are statistically preferred paths, that is, the average electron will, on a statistical basis, descend on the energy ladder by a specific path. In the case of gases, this results in narrow bands of emitted energy. Following are examples of elements that in gaseous form emit most energy in a few narrow bands:

Element	Wavelength of Most Significant Emission (nm)	Apparent Color
Sodium	589,590	Yellow
Lithium	610,670	Orange-red
Lead	406	Blue-violet
Barium	553, 614	Yellow-green

The resulting color appearances have been used in analytical chemistry to help identify materials. In the case of incandescent solid materials, quanta of individual atoms or molecules have more widely varying energy levels, resulting in continuous energy distributions.

The amount and energy distribution of emitted light are functions of the temperature of the emitting matter. The higher the temperature, the higher the amount and average energy level of the emitted quanta. Emission ceases completely only in the vicinity of the lowest possible temperature, that is, 0 kelvin (4). To be seen as light, the temperature of the emitting material must be above 1000 K.

BLACKBODY RADIATION

A blackbody is an idealized nonexistent material that is a perfect absorber and emitter of energy. It absorbs and emits energy indiscriminately at all wavelengths. At a given temperature the emission of such matter can be calculated on a theoretical basis. Examples of black body emission at different temperatures are illustrated in Figure 1.2. Many real materials produce an emission spectrum quite similar to that of a blackbody. Blackbody temperature, expressed on the absolute Kelvin scale, is in turn routinely used to qualitatively express the emission behavior of a light source even if its emission spectrum is unlike that of a blackbody. Thus, light sources are classified by their correlated color temperature, that is, the temperature of a radiating blackbody that has the same apparent color. Figure 1.2 also indicates that the emission spectrum of the sun as measured on earth quite closely resembles that of a blackbody at approximately 6000 K. It also shows that brightness sensitivity of the human visual system is tuned to the emission of the solar spectrum.

Returning to our example of a blacksmith and stating that, at least at higher temperatures, the emission spectrum of iron is close to that of a blackbody, the apparent change in color at increasing temperatures can now be explained in terms of the emission spectrum, as illustrated in Figure 1.2. The burning coal surrounding the metal radiates like a blackbody at a temperature of about 1800 K. At this temperature the emission in the visible range is low at low wavelength and high at high wavelength. Such a spectral power distribution is commensurate with light having an orange-reddish appearance. The common incandescent light bulb (in which a tungsten wire is made to glow by its resistance to the flow of electrical current) also has an emission spectrum close to that of a blackbody. Incandescent lamps are typically operated at 2500 K, with an approximate emission spectrum as illustrated in Figure 1.2. It is evident that an incandescent lamp does not make efficient use of energy, since most of the emitted radiation is not visible. Incandescent lamps become very hot during operation because most of the emitted energy is in the infrared region, and we sense that energy as heat. Fluorescent lamps, on the other hand, emit most of their energy in the visible spectrum and thereby operate cooler and are more energy efficient. The most energy-efficient fluorescent lamps are the so-called triband lamps emitting light in three relatively distinct bands around 440 nm, 540 nm, and 610 nm. Because in the other regions of the visible spectrum their emissions are low, they are more energy efficient than other fluorescent lamps that emit light throughout the whole visible

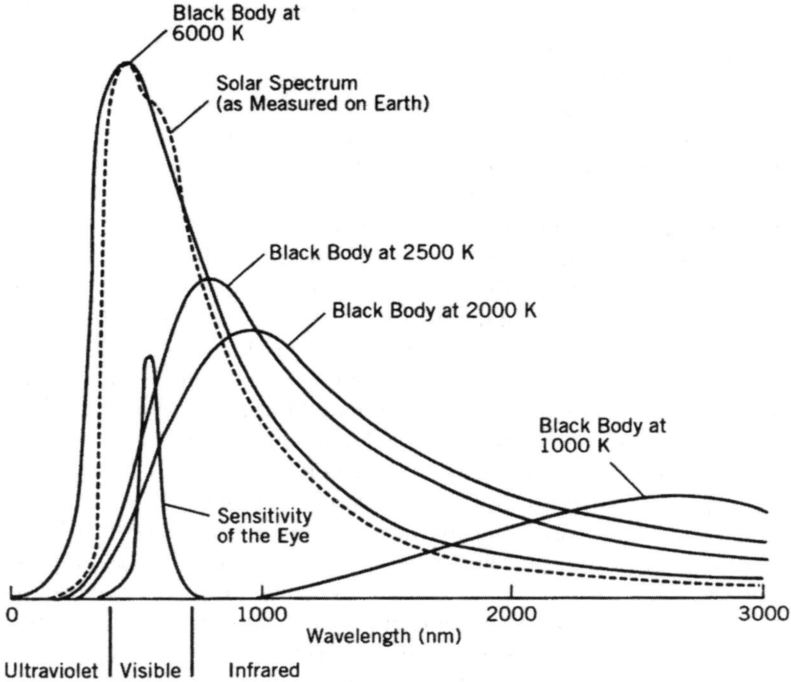

FIGURE 1.2 *Blackbody emission spectra at various temperatures (in degrees kelvin), the solar spectrum as measured on the surface of the earth (dashed line), and the spectral brightness sensitivity of the human visual system.*

range. The appearance of certain reflecting materials can change significantly as a function of the spectral power distribution of the light under which they are viewed (see color constancy, Chapter 4).

Blackbodies at temperatures beginning at 2500 K and higher emit light that, especially after adaptation (see Chapter 3), is seen as colorless. When objects with a high flat reflectance function are seen in this light they appear white. As a result such light is commonly termed *white.* This neutral experience is our response to the pervasive presence of daylight in our life. There are many other spectral power distributions that result in the corresponding light appearing colorless, or "white." One thing they all have in common is that, despite their variation in spectral power, they have an effect on our visual apparatus very similar to that of daylight.

LUMINESCENCE

Light also can be created by processes not based on the absorption of energy. This phenomenon is called *luminescence*. There are three basic processes: electroluminescence, chemiluminescence, and photoluminescence.

Sparks, arcs of light, lightning, some types of laser light, and gas discharges are examples of electroluminescence. Here, under the influence of an electric field, electrons collide with particles of matter, resulting in the emission of the appropriate energy level to be seen by us as light. Chemiluminescence is produced at low temperatures by certain chemical reactions, mainly oxidations. Natural chemiluminescence, also called *bioluminenscence*, can be observed in glowworm, fireflies, and certain deep-sea fish, as well as on decaying wood or putrefying meat. Glowing liquid-filled plastic tubes are a commercial form of chemiluminescence.

Photoluminescence appears in two forms: fluorescence and phosphorescence. Fluorescence is due to the properties that certain molecules have to absorb near-ultraviolet or visible light and shed it not in the form of infrared energy, as most absorbers of visible energy do, but in the form of visible radiation of a somewhat higher wavelength (that is, lower energy content). Fluorescent whitening agents, present in many detergents, absorb ultraviolet radiation between 300 and 380 nm and emit visible radiation from 400 nm to 480 nm. This light has a bluish appearance, and materials treated with such products appear very white in color. Fluorescent dyes or pigments (see also Chapter 8) absorb and emit visible energy, for example, a fluorescent "red" dye absorbs light from about 450 nm to 550 nm and emits light at 600 to 700 nm. Fluorescent colorants appear to glow faintly because of the emission of light, but they are weak emitters. There are also inorganic materials that fluoresce, for example, certain minerals. Fluorescent light tubes are another example of the process of fluorescence. The tubes are coated on their interior with fluorescing phosphor compounds. They contain a small amount of mercury that is brought to the incandescent state with the application of an electric field. The energy emitted by the mercury is in the near-ultraviolet. It is absorbed by the phosphor compounds that in turn emit broadband visible light. The term *fluorescence* is applied in cases where the emission of light stops at the same time the flow of absorbed energy is interrupted. Some substances, for example elementary phosphor, are capable of storing absorbed energy for a time. They continue to emit light for some time after the exciting energy is interrupted. This process is named *phosphorescence*.

ABSORPTION, REFLECTION, SCATTERING, AND TRANSMISSION

From creation to oblivion the fate of light can pass through many stations. If it consists of a broad band of energy, selective action at different energy levels results in changes in the spectral power distribution, and when viewed may result in color experiences. When light quanta are absorbed by matter, that is, if the photons of the light beam interact with atoms or molecules that can respond to their energy level, the result is loss of energy by the quanta and later reemission at a lower energy level, typically in the infrared. The radiation is lost as a visible stimulus and has become a stimulus sensed as heat.

By definition, the most efficient absorber is the blackbody, which absorbs and emits energy indiscriminately (if by strict rules) over a wide energy band. Real objects are often selective absorbers. Of particular interest in this discussion is their absorption

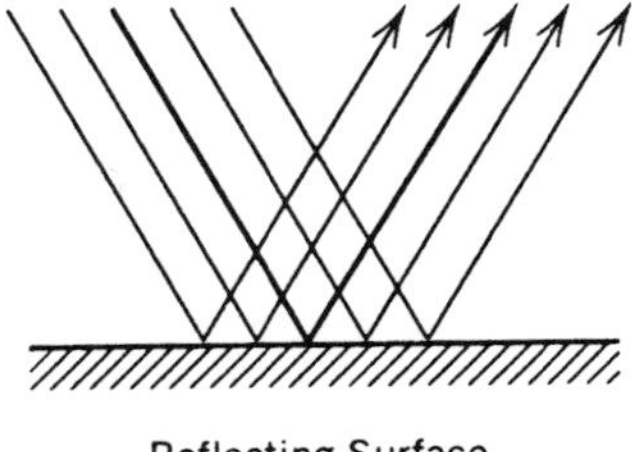

FIGURE 1.3 *Reflection of light from a plane surface.*

of visible light. Some absorb very little, say, a layer of "white" paint, a lot, such as a layer of "black" paint, or at any level between. Real objects do not absorb all light energy falling on them, and some of the photons are scattered or reflected. Reflection is a special form of scattering. It is the process by which photons arriving at a smooth-surfaced material change their direction of travel on impact and are returned (like a ball thrown against a wall) (5). In the case of reflection, the angle of incidence (the angle at which the photons strike the surface) is equal to the angle of reflection (Fig. 1.3). Reflection is unequivocally predictable, while scattering is only predictable in a statistical sense. Scattering refers to the change in direction suffered by radiation on impact with a rough-surfaced material or with fine particles of uniform or varying shape. In this case, reflection is in many directions. The surface involved may appear smooth to our senses, as does the surface of a dried layer of paint. However, the pigment particles in the paint form a microscopically rough surface, scattering light in many directions (Fig. 1.4). Typical scattering materials are textile fibers (small diameter, comparatively smooth columns of matter); water droplets suspended in air in the form of clouds or fog; smog and dust particles; milk (fine oily droplets in a water-based emulsion); and some types of bird feather, for example, those of blue jays. Many colorants, particularly pigments, are scattering materials. Many artificial materials display a complex interplay of external reflection, transmission, and internal scattering of light, for example, glossy paint.

Scattering of photons occurs in the atmosphere as a result of water droplets, ice crystals, or dust particles. Without it the sun's light would be very harsh in an otherwise

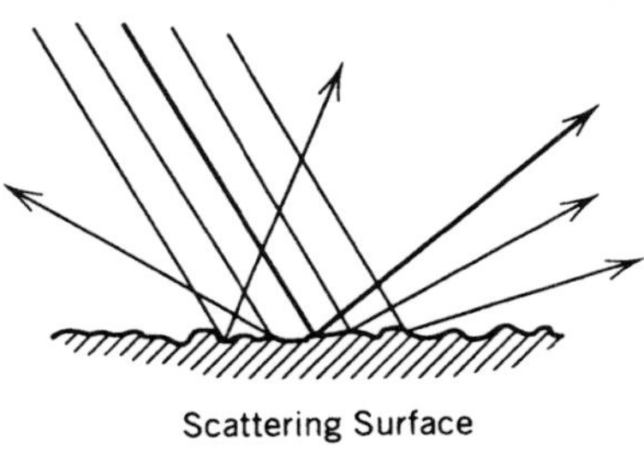

FIGURE 1.4 *Scattering of light on an uneven surface.*

black sky, such as astronauts on the moon have experienced. Scattering causes the diffused daylight we experience on the surface of the earth. Such scattering is dependent on the size of the particles in the air and wavelength of light. Larger particles or a high density (such as in a fog) scatter all light equally and are perceived as white. Heavily scattered sunlight, such as on a very cloudy day, in fog, or a snowstorm, seems to have no origin: photons meet our eyes from all angles and shadows are soft or nonexistent.

Few and small particles scatter short-wave light rays more efficiently than long-wave rays. While most rays of longer wavelength pass through the atmosphere unscattered, a higher proportion of short-wave light is scattered, resulting in a blue appearance of the clear sky. Clouds, consisting of water droplets or ice crystals, scatter light of all visible wavelengths equally and appear white. The chance of a photon being scattered also depends on the thickness of the layer it passes through. Thus, near sunset, and especially in an atmosphere with high amounts of particles (for example, in an industrial area, or after a volcanic eruption), all light except that of the longest wavelengths is scattered, causing the sun's disk to appear red. As mentioned, the blue appearance of the feathers of birds like blue jays and kingfishers are also caused by scattering at their surface.

Perfectly reflecting or scattering materials do not exist. Some come quite close, for example, a pressed surface of pure barium sulfate scatters some 98% of photons in the visible region of the spectrum. Some of the best reflecting materials are metallic mirrors. They reflect 70 to 80% of photons arriving at their surface.

Most color stimuli we encounter are the result of wavelength-specific absorption and scattering. They are known as *object colors*. They absorb or scatter all visible wavelengths to a greater or smaller degree. Figure 1.5 represents the spectral reflectance function of an object seen as having a blue color when viewed in standard conditions. Reflectance curves represent at each wavelength the ratio of the numbers

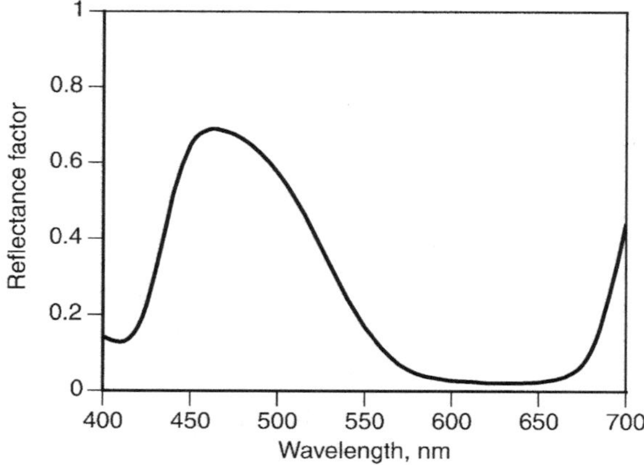

FIGURE 1.5 *Spectral reflectance function of an object causing a perception of blue when viewed in standard conditions.*

of photons leaving the surface to that arriving at the surface of the object (see Chapter 6 for further discussion).

Transmission refers to the mostly unimpeded passage of light through a transparent object, such as a layer of pure water. The spectral distribution of a light beam, after passing through such a layer, is unchanged. If the layer contains absorbing materials (e.g., dyes), a portion of the light is absorbed and the remainder transmitted. The amount of absorbed light in this case depends on the thickness of the layer. If the dye is completely dissolved (at the molecular level, that is, free of agglomerates causing scattering), Lambert's and Beer's laws allow the amount of light absorbed and transmitted to be determined (6). The size of single molecules relative to the wavelengths of light defines if a material, completely dissolved, results in scattering.

REFRACTION

The term *refraction* is used to denote a change in the direction of a stream of photons when passing from one medium into another. When light that is passing through air obliquely strikes the surface of a transparent object, such as water or glass, it changes direction according to the laws of refraction (7). This phenomenon is the basis of the rainbow or the image formation in a camera or in the eye. In both camera and eye, refraction is controlled by lens design. Photons striking the surface of a photographic lens or the lens of an eye at a given position (except for the center) change direction as they pass through the lens and exit it, and in this manner are focused on the film or the light-sensitive layer of the eye, forming an image, inverted and reduced in size, of the world in front of the lens. The change in direction is a function of the optical densities of the two transparent media involved (lens material and air) and of the energy level of the photons (that is, their wavelength). Photons of higher energy change direction more strongly than those of lower energy. Refraction, therefore, is an effective technique for separating the components of a mixture of wavelengths, such as sunlight. A glass prism is a useful practical tool to accomplish this: when a narrow beam of polychromatic light passes through a prism, its components are separated as they leave the prism (see Fig. 1.6). The individual components when seen reflected from a "white" screen, are perceived to be colored. If the light used is daylight, the perceived colors are those of the complete visible spectrum. Light from (always approximately) 400 nm to 490 nm causes a bluish experience, from 490 nm to 570 nm a greenish, from 570 nm to 590 nm a yellowish, from 590 nm to 630 nm an orangeish, and from 630 nm to 700 nm a reddish experience. When the direction of the flow of photons is reversed, the resulting stimulus, when viewed under standard conditions, is seen as white again.

The most spectacular natural display of refraction is the rainbow. Refraction effects can also be seen in cut crystal, diamonds, or other gemstones having "fire." A difficulty resulting from refraction in lenses is known as *chromatic aberration*. Because of the specific effect of refraction on light of different wavelengths, its photons emanating

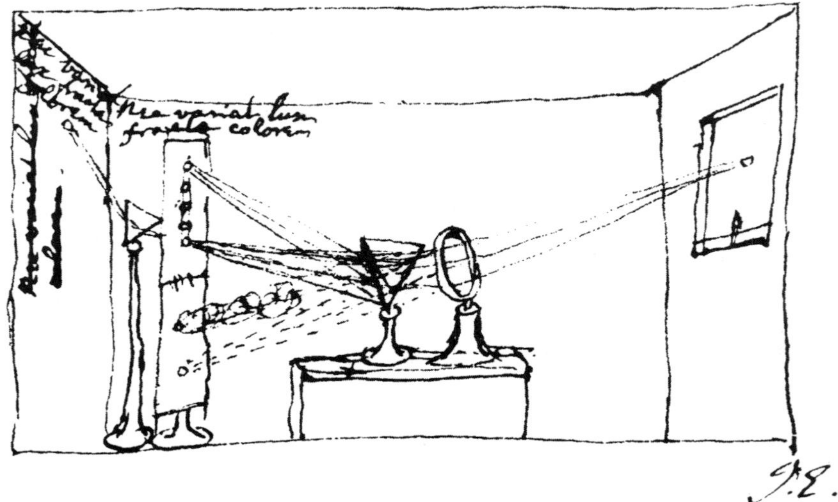

FIGURE 1.6 *Newton's sketch of his experiment where he used a prism to refract sunlight into its spectral components. The light passing through a small hole on the right is collected by a lens and passes through the prism where it is refracted into its spectral components in an elongated band on the upper left. Openings in the screen allow light of certain wavelengths to pass through. A second prism behind one of the openings shows that the refracted-narrow band light coming from the first prism does not change further in passage through a second one. In a separate experiment he also showed that this process is reversible.*

from a given point and passing through the lens can only be focused on a common point on the other side of the lens if the lens has been corrected for chromatic aberration.

INTERFERENCE

Puddles of water with bright multicolored bands on the surface are a common occurrence near a car repair shop or a gas station after a rain shower. Similarly, and esthetically more appealing, bright shimmering colors can be seen on the wings of some butterflies when viewed from a certain angle, or on the feathers of some birds, such as the peacock or some kinds of hummingbirds. Such colors are called *iridescent* and differ from the scatter-effect colors of the blue jay. Hue and intensity of color appearance change with the angle at which the surface is viewed. These colors are due to a physical effect called *interference*, a term used to denote the temporary splitting of light waves into parts that are later recombined. Depending on the path the beam components follow after splitting, the light waves may be in or out of phase when recombined, that is, the wave peaks and valleys may or may not match. If they do match, the intensity of the resulting beam is the sum of those of its components; if they do not match, the two components cancel each other. A typical source of interference is a thin transparent film, such as an oil film on water or a soap bubble. Whether or not the reflected light will be in or out of phase depends on the thickness of the film. If in phase, light of varying wavelengths will emerge at corresponding angles, giving rise to pure, strong color stimuli, the color of which depends on the angle of viewing. Several colors (as in a thin oil film on the surface of water) can be seen if the film causing interference varies in thickness.

DIFFRACTION

Diffraction is a special case of the combined effect of scattering and interference. The behavior of a light wave arriving at the edge of a solid material (think of the edge of a razor blade) is influenced by the sharpness of the edge. Depending on its wavelength in relation to the dimensional properties of the edge, the wave either passes unimpeded by the edge, is scattered at the edge, or is absorbed, reflected, or refracted by the edge-forming material. If several properly dimensioned edges exist, such as when fine lines are inscribed or etched into a glass or metal plate, the resulting scatter at the edges is subject to interference effects: waves in phase will reinforce, while those out of phase will cancel each other. When daylight strikes such an assembly of edges (called a *grating*) waves in phase are enhanced in different directions: a display of spectral colors results when viewed from different angles. A typical example is the surface of a compact disk, although because of the irregularity of the embedded digital patterns and their curvature the effect is less than perfect. Gratings made by an inscribing process, called *ruling*, or other techniques are used widely today in optical equipment for separating polychromatic light into its components.

Certain organic substances, such as the wings or body parts of some insects, have structures with the dimensions necessary for diffraction effects. Liquid-crystal molecules represent another example. They are arranged in a crystalline configuration such that they act as diffraction gratings. The dimension of the edges is a function of the surrounding temperature, and such devices can be used as temperature indicators, among other applications. The colors of the gem opal are also a result of diffraction.

MOLECULAR ORBITALS

So far the concern has been about physical sources of color stimuli (refraction, interference, etc.) and about the behavior of excited electrons in atoms and molecules. In atoms as well as in molecules electrons are arranged in orbits around the nucleus. When electrons in the outermost orbit, called *orbital electrons*, from two atoms form a stable pair, the result is a chemical bond and formation of a molecule. In some molecules, orbital electrons are not confined to a particular location, but can range across larger areas. Such behavior can give rise to color stimuli. A typical example is the gem sapphire, the basic material of which is aluminum oxide, capable of forming crystalline structures. In its pure form, called *corundum*, aluminum oxide is not a source of a color stimulus. Sapphire contains a degree of impurity in the form of iron and titanium atoms replacing aluminum in some of the molecules. Ionized aluminum

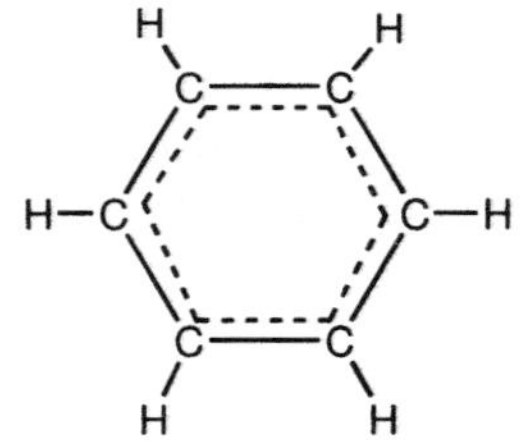

FIGURE 1.7 *Structural formula of the organic molecule benzene. C represents carbon, and H represents hydrogen atoms. The dashed line represents three electrons of variable location.*

has three electric charges, while iron has two, and titanium four. One of the electrons from titanium tends to transfer to a neighboring molecule containing iron. As a result, both atoms end up with three electrons. This charge transfer, resulting in an excited state of the electrons, occurs only under the influence of absorbed energy. The needed energy can be supplied by absorbed photons of the visible range in a broad band from approximately 550 nm to above 700 nm. The energy released by the excited electrons is in the infrared band and therefore not visible. As a result, only light from 400 to 550 nm is reflected, resulting in a deep-blue color sensation.

A somewhat similar process takes place in most dyes and organic pigments. They consist of organic molecules made up mostly of carbon, oxygen, hydrogen, and nitrogen. Carbon atoms (as well as, under certain circumstances, those of nitrogen) can bond with other carbon atoms and form chains with alternating single and double bonds. The best known example is the closed chain, or ring, of the benzene molecule, the carbon chain of which consists of six carbon atoms with nine electron bonds (see Fig. 1.7). Depictions of the structure of this molecule usually show alternating single and double bonds. However, the orbital electrons are not located in a specific place, but range over the total ring. Such bonds are said to be *conjugated* (8). Benzene absorbs light in the ultraviolet region. In other, more complex molecules containing benzene rings the absorbed energy is often of the visible range and the substances appear colored. Molecules containing this kind of conjugated bonds are called *chromophores* (color bearers).

It is possible to attach to these molecules side groups capable of accepting or donating an electron. Such groups are called *auxochromes* (color increasers), and they affect the absorption characteristics of the chromophore to which they are attached. Two well-known natural substances derive their color from conjugated bond systems: blood and chlorophyll, the life-supporting substances of animals and plants, respectively. Most organic colorants represent synthetic examples. It is likely that in the last 200 years hundreds of thousands of different molecules with conjugated bond systems have been synthesized in laboratories around the world in a never-ending search for better colorants. Some 8000 of these have found commercial significance in the past or have it today and are listed in the *Color Index* (9). For a more detailed discussion of colorants, see Chapter 8.

Fluorescent colorants (also discussed in the "Luminescence" section) represent a special case of conjugated-bond systems. The wavelength of the absorbed energy is in the near UV or the short-wave visible range. The emitted energy is not in the infrared as usual, but in the visible range, the shift being usually only some 100 nm. Chromatic fluorescent colorants are both absorbers and emitters of visible energy.

CRYSTAL-FIELD COLORS

A dramatic display of crystal-field color can be witnessed by visiting Bryce Canyon National Park in Utah with its huge, Gothic, weather-beaten columns of reddish sandstone. The sensation of red is caused by iron oxide present in the sandstone. Other well-known examples of crystal-field color are the gems ruby and emerald. Ruby, like sapphire, is based on aluminum oxide, but with some aluminum atoms replaced by chromium. The crystalline structure of aluminum oxide creates an electric field, the crystal field. The chromium atoms present result in absorption of the middle portion of the visible spectrum, and a ruby, when viewed in daylight, is seen as having a deep bluish red color due to reflection or transmission of the short- and long-wave bands of light. The red color is due not only to absorption but also to fluorescence. One step on the excitation ladder results in emission of long-wave light contributing to the ruby's redness. Color perceptions resulting from viewing colorations made with many inorganic pigments, such as iron oxide, and those from viewing semiprecious stones like turquoise derive from crystal fields.

ELECTRICAL CONDUCTORS AND SEMICONDUCTORS

In materials known as conductors and semiconductors, the orbital electrons are not restricted to the atoms of which they were an original part, but instead they can travel over the whole volume of the material. Thus, in copper wire some electrons can travel from one end of the wire to the other. On application of an electric field, the free electrons produce an electric current. Metals have what is known as a *plasma frequency*. Light of higher wavelengths (lower frequency) than the plasma frequency is reflected. Light of lower frequency can pass through metals, that is, the metal is transparent to such energy. For chrome, for example, the plasma frequency is in the ultraviolet region, making chrome a reflector of all visible wavelengths and thereby a good mirror. For copper the plasma frequency falls into the visible region, giving copper a reddish hue.

Semiconductors (the best-known example is silicon) have four electrons per atom or molecule capable of undergoing chemical bonding. Semiconductors have a gap in their energy absorption behavior: they can absorb energy only in bands separated by a gap. Depending on the position of the gap in the energy ladder, the result can be absorption of visible energy, and thereby generation of color stimuli. One well-known color semiconductor is the pigment cinnabar (mercury sulfide). Another is cadmium sulfide, used as a yellow pigment. Color stimuli from some of the semiconductors derive from inclusion of impurities by a process called *doping*. Such products are used in light-emitting diodes (LED displays of electronic gadgets). The value of some semiconductors does not lie in their color-producing capabilities, but in how they conduct electricity: they are important in electronics technology.

The list of sources of color stimuli described so far is not exhaustive and the reader is encouraged to explore further (10). In all cases (except the optical/geometric ones), the outer electrons in atoms and molecules and their movement on the energy ladder are responsible for the creation of photons that act as color stimuli. There is a significant level of scientific understanding of the causes of color. But it is understanding at an intermediate rather than truly fundamental level, because it is limited by issues and difficulties in quantum theory. It is thorough enough now to make it possible to create on demand synthetic substances of very specific absorption and emission behavior. An expressive example of the level of understanding is the development of the laser (11), a device to create intense beams of light of specific wavelengths. The light consists of photons of identical wavelength with all waves being in phase, that is, in step. Such light beams are termed *coherent*. Ordinary light, (e.g., daylight) is a composite of photons of many wavelengths traveling partly in phase but mostly out of phase.

The first laser used ruby as a lasing material. In response to the light from a flashbulb, the outer electrons of the chromium impurity in ruby are raised to a highly excited state. Photons emitted by the electrons, on returning to their normal state, result in an intense, coherent pulse of light perceived as red. With proper manipulation it is possible to generate a continuous beam of light. Other lasing substances are dyes in which, as seen before, absorption is created by conjugated bonds. Today it is possible to have laser beams of any visible wavelength, with many outside the visible region. Lasers have many practical applications, for example, as a cutting tool in manufacturing or in medicine (by precise focusing of an intense beam of energy resulting in heating, melting, evaporation, or combustion), in measuring of distances, or in holography. Their esthetic properties are put to use in light shows.

Today there are few if any color phenomena whose causes are not understood at the intermediate level of scientific understanding discussed earlier. The variety of color perceptions and the complexity of their nature help us to understand the difficulties over many centuries in understanding them (see Chapter 10). Sources of color perceptions outside of our bodies are all identical in that they cause photons of specific wavelengths to travel to our eyes. In the eyes their energy is transformed in complex ways to ultimately result in unconscious and conscious responses of the mind and body.

2

What Is Color and How did we Come to Experience It?

Many readers will be surprised to learn that as yet there is no scientific explanation of the color vision process, or the vision process in general. Light passing through the lenses of our eyes is absorbed by the two-dimensional, curved layer of light-sensitive cells in the retina (see Chapter 3). From this image the brain and mind somehow create the visual world in front of us that normally sighted people experience. There are ideas of how this might work, but none that is all-encompassing and generally found to be valid. Much has been learned since the early 1980s about the neurophysiological processes involved in color vision. But there continues to be a black box in our brain into which biologically produced signals disappear and out of which color experiences appear.

Like all other senses and functions of the body, the visual sense has many aspects. There is the comparatively well-known anatomical aspect, the eye with its parts and the strands of nerve fibers from eyes to different regions of the brain. There is the physical/optical aspect: the description of light, its interaction with objects, transmission through and absorption by various materials making up the eye, and the fate of the light trapped in the eye. There is the physiological aspect: the biological processes of the visual system: transformation of physical light energy into chemical energy and its transmission within the brain. There is finally the psychological aspect: the response of the organism to the absorbed energy expressing itself in behaviors and actions. The conscious psychological aspect includes the experience of qualia, the designation given to perceptual conscious experiences like color, taste, sound, that are qualitative indicators of materials (1).

Color: *An Introduction to Practice and Principles, Second Edition,* by Rolf G. Kuehni
ISBN 0471-66006-X Copyright © 2005 John Wiley & Sons, Inc.

The continuing mystery of consciousness is only one aspect of the problem of vision. We "see" many more things than what we are conscious of. Under daylight conditions our eye/brain system processes some 10 million bits of information per second, of which we can only process some 40 bits per second in consciousness (2). The brain acts as a huge filtering device to screen out much of the data. Of the processed information the major portion is used in subconscious processing, with many effects on internal activities and movements we are not aware of or only after they have happened. Only a small portion reaches our consciousness. Surprising results have been obtained in recent years in so-called change blindness experiments where an image is replaced by a second identical looking one in which, however, something significant has been changed. Of a group of observers up to 80% cannot say what has changed, even on repeated exposure (some up to dozens of times). In a film sequence in which viewers typically concentrate on the action, the fact that the color of an object not involved in the action slowly changes, say, from red to blue, is not noticed by most viewers (3). On the other hand, information that we subconsciously gathered in our visual field can later be used to advantage in a search task. A growing number of scientists and philosophers believe that the world we experience is an illusion of which colors are an important part.

What is color for? As autonomous, widely ranging individuals in a web of organisms our key task is to operate successfully in the world. To do that most of us must safely grow to reproductive maturity, acquire food, shelter, defend ourselves when necessary, locate a suitable partner, and raise offspring to a level where they can successfully operate by themselves. We could not achieve these tasks without acquiring information from our surroundings. For this purpose we have various kinds of sensors responding to different kinds of energy (or information) input. From these inputs our brain/mind computes images of the world and unconscious or conscious strategies for achieving our goals. Among the most basic tasks is to be able to distinguish between objects, close up or from a considerable distance. To a significant degree this is possible with black and white vision only. But as Figure 2.1 illustrates, color vision offers further strong advantages. It helps us to understand where we are, if our situation is safe or dangerous, if an apparent source of food is one we perused in the past without adverse effects, and in many other situations. But it does much more, as we know. There are nature, the arts, fashion, living spaces, entertainment, and so forth, where we profit from our ability to experience colors. These may be fortuitous by-products of the basic advantage we reap in distinction, but they are nevertheless a very important part of our lives.

As discussed in greater detail in Chapter 3, human color vision is mediated by three types of light-sensitive cells in the eyes, called *cones*. They are designated as *L, M*, and *S* (for long, medium, and short) depending on their spectral range of sensitivity. In these cells light energy is converted to chemical energy transmitted to various regions of the brain for further processing.

We are, of course, not the only organisms with color vision: some insects and birds, some aquatic animals, and many other mammals have kinds of color vision. Some of these are simpler, but others may be more complex than ours. We have three kinds of cones, as do bees (but at different wavelength ranges), many mammals only

TABLE 2.1 Evolutionary Timetable

Period	Time	Development
	4 BYA	Origin of life on earth
Proterozoic	2.5 BYA	Halobacteria with "vision," **development of rods with rhodopsin**
	2 BYA	Oxygen begins to accumulate in the atmosphere
	800 MYA	Plants and soft-bodied animals develop in oceans, **early form of *L* cone develops**
Cambric	500 MYA	Cambric explosion of animal types, hard-bodied animals develop, many with eyes
Ordovician	450 MYA	Plants and animals move ashore, ozone shield develops
Carboniferous	320 MYA	Pangaea supercontinent, forest growth
Permian	250 MYA	Reptiles develop, **separate *L* and *S* cones, the "ancient system," develops**
Triassic	230 MYA	Dinosaurs proliferate
Cretaceous	135 MYA	First mammals, flying reptiles, Pangaea begins to split
	65 MYA	Dinosaur extinction, first small primates, Africa separates from South America
Tertiary	35 MYA	Whales develop; ***L, M, S* trichromatic system develops in primates in Africa**
	3.5 MYA	First hominids in Africa
Quaternary	1.8 MYA	Homo erectus, our upright walking ancestor, appears
	100 TYA	First Homo sapiens, humans of the modern type, leave Africa
	30 TYA	Neanderthal-type people extinct, first cave paintings in Europe
	6 TYA	First writing

have two, but there is a type of shrimp that has nine different kinds of light-sensitive cells. The number of cone types and complexity of the color vision system appears to depend on the ecological niche in which ancestors of the organisms operated.

Based on DNA analysis geneticists have developed a plausible timeline for the development of vision in general and human color vision. The approximate timeline of major development steps is shown in Table 2.1.

Primitive vision, presumably because of its importance for survival, developed very early in the history of life. Rhodopsin was one of the first biological molecules reacting to absorption of light by changing its form. It has survived in many animals into the present time, including man. It is the basis of our night vision system involving the type of cell called *rod* (see Chapter 3). An early form of a cone, a different kind of cell functioning best at higher levels of light, developed about 800 million years ago (MYA) when life only existed in water. Movement of animals onto land and the new challenges connected with this change resulted in the development of a second cone type, the *S* cone. Together they form what is known as the *ancient system*. Rods as well as *L* and *S* cones use the same basic molecule, retinal, as light absorber. The difference in their wavelength sensitivity is due to different protein molecules to which retinal is attached. As a result, animals with such a system can distinguish between many more objects reflecting light between about 400 and 700 nm. At that time most of the Western world was connected into a huge continent geologists named Pangaea. In the Tertiary period tectonic activity began to split Pangaea into pieces that drifted apart. Important for our story is the splitting of the South American from the African continent. As a result of this split, all but one of the monkey species of the New World have two cones, while in the Old World (specifically Africa), some 35 MYA, evolving primates developed a trichromatic system with three cone types. Humans share trichromatic vision with these primates (4).

What was the cause for this development? The leading hypothesis involves food (5). While the dichromatic system allows animals to distinguish long from short wavelengths, the distinction in the middle region is comparatively poor. As a result animals with this kind of system have difficulty distinguishing between some greenish and reddish objects. Ripening fruit of many kinds change from green to yellow, orange, or red (see Fig. 2.1 for a general example). Also, the reddish color of young and succulent edible leafs in certain African plants changes to green when they are fully developed

FIGURE 2.1 *A* Cornus kousa *tree with fruit. Left: black-and-white image; right: identical image in color.* Figure also appears in color figure section.

and more difficult to digest. In both cases, an effective distinction between middle and long wavelength light was important for locating such food sources. Of course, we cannot assume that these early primates, or hummingbirds and chickens (having four cone types), experienced these wave bands as we do, that is, consciously as colors.

The machinery of seeing has apparently evolved multiple times. There are at least five major types of eyes, of which the two most important are the compound type (of most insects) and the camera type (of mammals and other animals) (6). The complete range of wavelengths to which these eyes are sensitive extends from near ultraviolet to near infrared, from about 300 nm to about 800 nm. Animals with sensitivity in the near UV are bees, pigeons, and roach fish. As seen in Chapter 1, this is a narrow range within the total electromagnetic spectrum. But it is the range produced in abundance by our sun, reflected or transmitted by most materials. Its absorption causes minimal damage to the cells involved (on prolonged exposure, sunburn), and it can penetrate water to a depth of some 20 feet, depending on how clear it is. The highly sensitive rhodopsin developed at a time when life existed in water only. Among land creatures it is found as an exclusive sensor cell type in nocturnal animals such as opossums and tarsiers. We do not have sensory sensitivity (at least that we know of) to any other range of the total electromagnetic spectrum.

THE OPPONENT COLOR SYSTEM

The sensitivity of the three cone types peaks at different wavelengths (approximately 440, 540, and 570 nm), but their light absorption functions overlap widely, particularly those of the *M* and *L* cones (see Fig. 3.7). Much of the information gathered by the two cone types is therefore duplicated, not an efficient mechanism. Narrower absorption functions may not be biochemically possible, and more kinds of cones would result in less visual acuity, that is, lower resolution of the image (comparable to fewer pixels per unit area in a digital camera). The problem was how to exploit to the fullest the information obtained from the three detector types. Nature solved it by developing an opponent color system, where the output of one cone type is subtracted from that of others. This allows for normalization of cone-derived signals in a way that makes distinctions according to three attributes possible. Further discussion is found in Chapter 3.

In principle, three detectors could have resulted in three pairs of chromatic opponent color systems. The fact that we only have two may indicate there was no evolutionarily significant need for further, more detailed discrimination capability, as is presumably available to other species equipped with tetrachromacy or even more different kinds of detectors. In recent years geneticists have found that about half of human females have the genetic potential for four cones (7). It is not known to what extent tetrachromacy exists in the female population and what, if any, effect it would have on color perception.

There are interesting theories about coevolution of color vision capabilities in insects and birds and development of colors in flowers and fruit. Such coevolution would have been beneficial for both animals and plants (8). Color, together with form, also began to be used as a tool of sexual selection.

WHAT IS COLOR?

On a simple level color is first and foremost a personal experience. It is an important part of the total visual and emotional experience we have, for example, when standing at the edge of the Grand Canyon at sunset, or when walking through the galleries of the Uffizi museum in Florence. We can identify without difficulty with the statement of the philosopher Bertrand Russell: "I know [a] color perfectly and completely when I see it . . . " (9). There is little if anything objective about color experiences; only each individual knows what she/he experiences. Any objective part of color deals entirely with the definition of color stimuli, but, as discussed in more detail in Chapter 4, there is considerable variation in color stimuli perceived as having unique hues, particularly for green and red. Selection of such stimuli is not directly tied to variation in the specific cone sensitivity of the observers and presumed resulting subtractive opponent color functions. Color perceptions seem to be generated at a level of brain activity beyond that of the generation of cone opponent signals.

Autonomous animals need to be able to assess situations in their surroundings and rapidly develop strategies to deal with them. Failure to do so successfully can mean death, and thereby failure to help continue the species. As a result, there must be mechanisms to take visual and other sensory clues and compute from them strategies that prove successful. For animals with simple brain structures such operations are believed to be performed automatically, without any conscious aspects. Some parts of the brain "experience" the opponent color signals and use the information to produce muscle movements and other behaviors: for example, a hummingbird flies in the direction of the colored flowers that promise nectar. For animals living in groups many new problems arise: how to interrelate among members, if and when to share food, if and when to join defensive or offensive alliances, when to stay and when to move on, and countless more. Brain structures to successfully deal with issues of this kind must be more complex and larger. At one point the creature no longer operates in robotic fashion, but in addition develops consciousness, self-awareness, and aware strategies (10).

There is considerable discussion going on today about where consciousness begins in the line of animals. In the past it was generally assumed that only humans are conscious, but most owners of pets such as dogs and cats, or people who have extensively interacted with primates such as chimpanzees and gorillas, find it difficult to deny them at least some degree of consciousness (it is important to keep in mind that no clear definition of the word exists). Consciousness means we can remember things from the past and can use such memories in consciously dealing with a given situation, a similar version of which we have experienced in the past. Even though many of our actions are automatic and instinctive and we perform them without knowing why, there are others where we can use memory and conscious cognition to come to a decision: The last time I ate a fruit of this shape, color, and smell I suffered a digestive upset; my sweetheart likes only pink roses, not white or yellow ones, as I learned in pleasant or painful past experiences.

Being able to contemplate the past requires having symbolic representations of some sort available for those earlier experiences, since one cannot reexperience them in their original form. In consciousness these symbolic representations must seamlessly connect past with present experiences. Over time the quantities of information representing conscious experiences are very large and their efficient storage and near-instant accessibility proved vital. The need for sorting out more important from less important memories and for high efficiency in storage resulted in condensed, symbolic representation of important information.

Without consciousness, as presumably for hummingbirds and bees, there is no aware memory of pink. There may be a neural network making them fly toward spots with a given spectral signature, genetically based and trained by actual experiences. In our case we have a conscious experience and memory of pink. It is represented by what is technically called a quale (plural: qualia). For us the quale of pink is something we can recall from past experience into consciousness, for comparison against what we currently see at the florist. Of course, we do not have to think about this any longer; it happens rapidly and seemingly automatically. We learn the name for pink early in our life and in whatever language we speak, but the quale itself presumably has a genetic basis. There may be a degree of commonality behind different kinds of qualia (colors, forms, tastes, sounds, smells), as the facts of synesthesia indicate (11). There is not full agreement among philosophers that qualia are a meaningful concept.

The nature of consciousness and that of such subjective qualities is unknown and a matter of great debate (12). Philosophers' opinions range from pan-psychism (everything in the universe is conscious, consciousness is a dimension of the universe in addition to space and time) via the idea that consciousness and qualia are illusions produced by the brain/mind to the thought that qualia do not exist (13).

As briefly mentioned in the previous chapter, one of the major philosophical discussions about color is if they are located in the outside world or in the brain/mind; that is, the argument between color subjectivists (internalists) and objectivists (externalists) (14). Many philosophers are, for what appear to be technical reasons, objectivists, that is, they believe color is located in materials and lights. But so far they have not been able to define what it is in the world around us that represents specific color experiences. There are several arguments on the side of the subjectivists, for example:

1. A given color perception can be caused by widely varying reflectance or spectral power functions (metamerism, see Chapter 4).
2. An object with a given reflectance function can result in significantly different color experiences depending on the white light source it is illuminated with (not even considering colored lights).
3. A given color sample is, under identical conditions, seen as having a distinctly yellowish green appearance by one observer, a distinctly bluish green one by another, and as a green neither tinted with yellow nor blue by a third (a spectral light identified as having a unique green hue by one observer may be

identified as having a unique yellow by another, both considered color normal; see Chapter 4).

4. Given color patches can result in distinctly differing color experiences, depending on the nature of surrounding patches (see Fig. 4.6).

Such effects appear to point to a process where a given spectral power distribution arriving at the eye is interpreted by a given brain/mind based on the context of spectral power distributions it is placed in and the interpretation of the scene the brain/mind arrives at. Such interpretations may be subjectively influenced by early vision experiences.

As mentioned earlier, color qualia can be interpreted as symbolic entities for the purpose of distinction between color stimuli. The symbolic entities of white, black, yellow, red, blue, and green have proved adequate to convey a large number of changes in stimuli at a given level of lightness. The number of discernible color experiences in a constant lightness plane has been estimated as 17,000, the total at all lightness levels in the millions (15). The process can be sketched as follows: The spectral information arriving at the retina is reduced to the output of three cone types in the retinal net, forming a two-dimensional map of the world in front of the eyes. The brain uses the result to compute an interpretation available for subconscious processing. Here, in animals equipped with the proper position of eyes, the information is translated into a three-dimensional map, allowing judgments of depth. Humans symbolically experience given aspects of patches in the field as colors to aid in their distinction from other patches. Accordingly, "blue" is the abstract symbol attached by the brain/mind to its interpretation of the radiation received from a given patch in the field of view, just as the taste of "sweet" is the symbol attached to the sensing of organic molecules with certain organic molecular structures, or the sound of "b flat" is the symbol for a certain vibratory state of the air mass surrounding our head.

Symbols can be more or less abstract, that is, have a more or less close relationship with some tangible aspect of reality. The symbol of a deer on a road sign is quite concrete in that it advises us that animals of that kind may be in the vicinity. In the case of colors, there are, so far, no clear-cut phenomenological relationships between colors and, say, reflectance properties of materials: there is no necessary connection we know of between the vibratory state of glowing sodium atoms and their reddish yellow color, and there is no necessary connection between the physics and chemistry of the surface of an apple and its apparent red color. The symbolic nature of color is supported by the fact that there are no evident (except familiarity) advantages to colors in a color photograph as compared to those in a geographical map. "False colors" of medical or astronomical images clearly indicate the value of colors as tools of discrimination, information carriers, similar to the real colors of our experiences. Colors, of course, have over millennia accumulated associations with cultural values. Such associations vary between cultures and within cultures from time period to time period. They have nothing to do with what color is. The old search of popular psychology for universal meanings and effects of colors has, therefore, been unsuccessful.

There is no disputing that spectral signatures of objects and lights importantly affect the brain/mind's decisions as to what symbolic color entity to apply to given regions identified as representing an object (or parts of one). But it is clear that there is no simple 1:1 relationship between such signatures and the resulting experiences. This would mean that only average observers, and only in a standard light and surroundings, experience the real color of an object, and all other experiences (that is, those of up to 90% of all human observers) are more or less false, including those of animals with a different number of cone types or different spectral ranges of these types.

To get back to the question in this chapter's title: at this time we do not know the answer to the question of what color is. We only have various speculative ideas, and there is no indication that this situation will change anytime soon. The next chapter describes the human visual apparatus in some more detail.

3

From Light to Color

Before describing the major color perception phenomena it is useful to delineate in some detail the apparatus of human color vision. Light enters our body through the eyes. The human eye (a schematic cross section is shown in Fig. 3.1) is an approximately egg shaped structure, held in place and moved by six muscles. It has a generic resemblance to a camera. Its shell, called *sclera*, is made of dense white fibrous material, except where it is usually exposed. There it is transparent and called *cornea*. The eye is filled with a transparent fluid named *vitreous humor*. Suspended in the vitreous humor and held in place by a system of muscle tissue is an elastic lens. Its shape is controlled by unconscious muscle action so as to focus through the lens an image of what is in front of the eye onto the inner back wall of the eye. When looking up from the pages of a book and gazing at the outside landscape, this system makes the necessary lens adjustments to project a continuously sharp picture of the outside world onto the inside back wall (complicated somewhat if the subject wears correcting glasses). Information from the surrounding world in the form of streams of photons passes through the cornea, the lens, and the vitreous humor to a slight indentation in the back wall, the fovea, on which most photons are focused.

The inside back wall is covered with a layer of light-sensing cells, the retina. Nerve fibers protruding from each cell form complicated webs, eventually forming the optic nerve. Between the light-sensing cells and their web of nerve fibers and the sclera membrane is another, highly pigmented membrane, the choroid. Its purpose is to absorb any photons not absorbed by any of the light-sensing cells.

The substances filling the internal space of the eye are not equally transparent to light of all wavelengths. A significant amount of shortwave radiation is absorbed

Color: *An Introduction to Practice and Principles, Second Edition*, by Rolf G. Kuehni
ISBN 0471-66006-X Copyright © 2005 John Wiley & Sons, Inc.

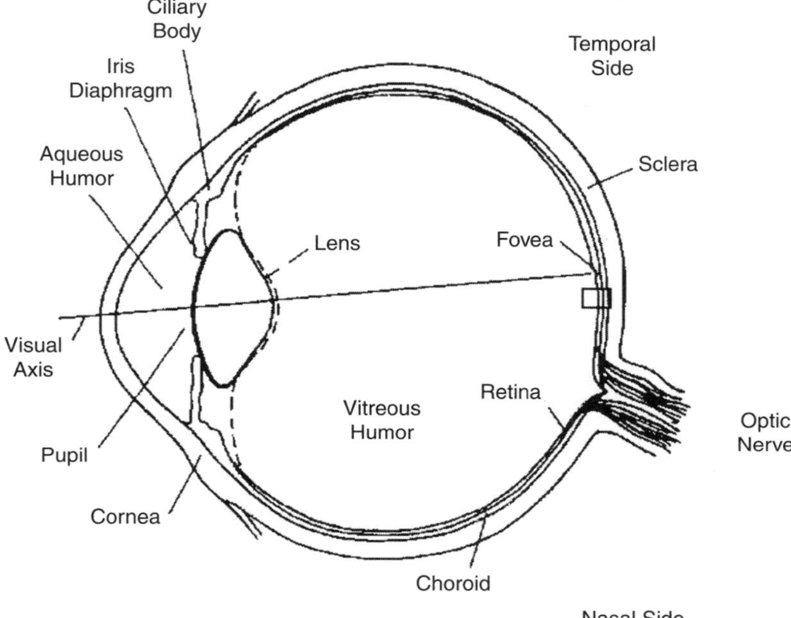

FIGURE 3.1 *Horizontal cross section through the human eye. The small rectangle indicates the location of Figure 3.2.*

by them. Absorption increases with age and as a result of certain illnesses. The number of photons entering the eye is controlled roughly by the circular opening of the iris. Incidentally, the color of the iris is not caused by colorants in the tissue but by diffraction effects. The size of the pupil opening is controlled by circular muscle tissue. The ratio of light passing through the pupil at its smallest and largest opening is about 1:15.

The retina contains about 120 million photosensitive receptors, some called *rods* and others *cones*. Their relative number varies in individuals and their distribution varies across the retina. There are only approximately 7 million cones, and they are located primarily in and near the fovea, the slight indentation in the retina at the endpoint of the visual axis. Rods are absent in the fovea. When moving away from the fovea the number of rods first increases sharply and then declines toward the edge of the retina (1).

Rods and cones are transducers in which the physical energy of photons is converted to electrochemical energy and passed along the optic nerve deep into the brain. Surprisingly, rods and cones do not face toward the lens but toward the choroid layer. Light therefore has to pass through the transparent retinal layer before being absorbed by rods or cones. A schematic illustration of the retinal layer is shown in Figure 3.2. Its real complexity is much larger than the figure indicates. While there are only two types of horizontal cells, there are about 10 different types of bipolar cells, the detailed

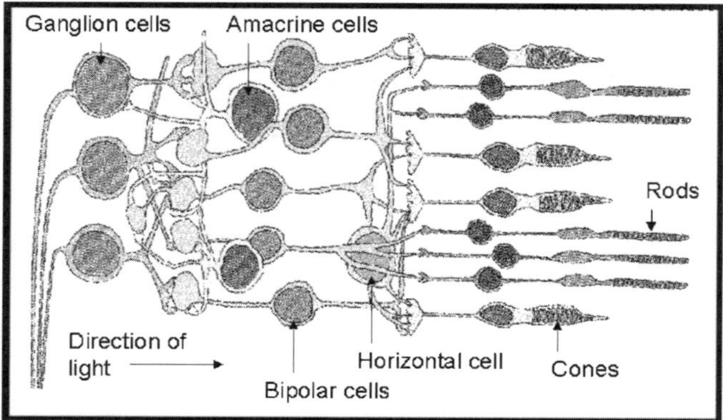

FIGURE 3.2 *Schematic representation of a cross section through the retinal layer of the eye.*

purpose of which is not yet known. Some bipolar cells are sensitive to structural elements such as borders and edges. There are now 25 different types of amacrine cells known. They interact with at least 10 types of ganglion cells. As a result, the structure is already hugely complex at this level. At the ganglion cell level the normal physiological activity of the cells ranges from spike firing at the resting state (approximately 10 spikes per second) to firing at an excited level (up to 400 spikes per second). These spikes carry the information generated by the cones. *S* cones and rods are connected differently from *L* and *M* cones. At this time, a reasonable consensus is that in *S*-connected ganglion cells the difference between absorption in the *S* and the sum of absorptions in the *L* and *M* cones is determined. In *L* and *M* connected ones their differences and sum are established. Some carry "center" information and others carry "surround" information of a group of cones that form a circular spot in the retina. In this respect ganglion cells can be separated into two groups: on-center and off-center. The former produces a spike when a spot of light falls onto the center of the receptive field of the cell, the latter produces a spike when the light falls on the periphery of the receptive field. This can be deduced by considering in Figure 3.2 the input from cone cells into a particular ganglion cell. The suitability of such cells to detect contours can easily be imagined. Receptive fields of neighboring lateral geniculate nuclei (LGN) cells are overlapping. The information passes along three major pathways out of the retina, designated as Mc, Pc, and Kc, where Mc stands for magnocellular, Pc for parvocellular, and Kc for koniocellular path. These proceed from the ganglion cells in the retina along the optic nerve to the LGN on the right and left side of the midbrain (Fig. 3.3). LGN are significant way stations on the path from the eye to the cortical visual centers in which additional computation of information takes place. Already in retinal cells beyond the cones the information no longer consists of absolute levels of photon absorptions, but comes from differences between outputs from various kinds of connected cells. The information in the magno and parvo paths ends up in different layers in the LGN, the former believed to carry basically brightness information, the latter two both brightness and spectrally opponent (color) information. Most of the

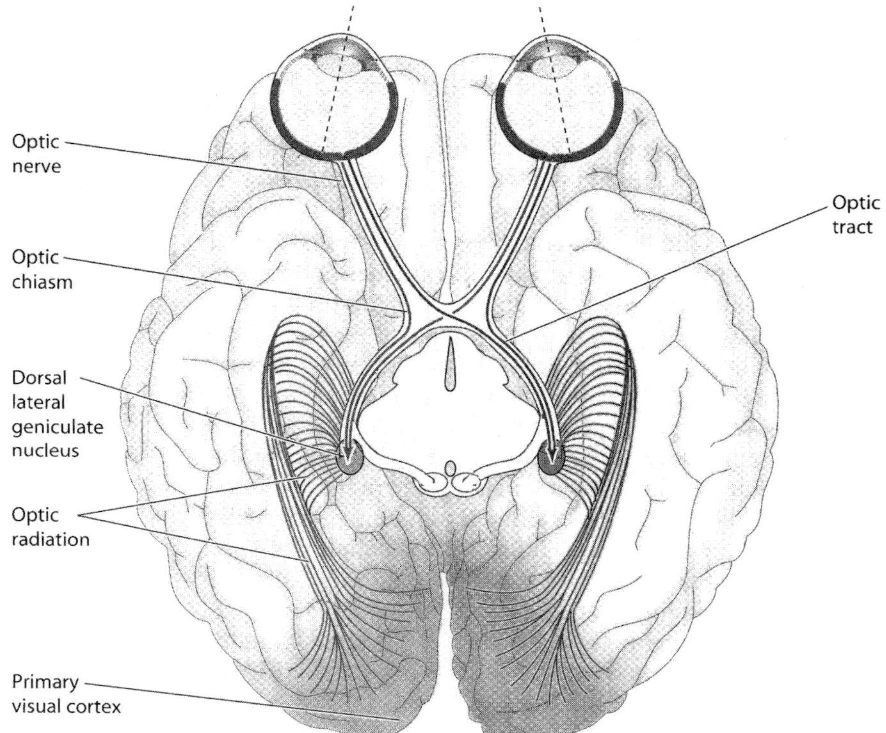

FIGURE 3.3 Horizontal cross section through the human brain showing the organization of the major visual pathways, beginning in the eye and ending in the cortical visual areas.

information is carried in the parvo path (Pc cells amount to some 80% of all visual cells) (2).

The exit of the optic nerve from the eye causes a circular blind spot in our field of vision. Normally, we are not aware of it, but it can be made visible. Before the optic nerves reach the LGN they pass through the optic chiasm where information from both eyes is separated in a manner that all information related to the left half of the visual field ends up in the right LGN and visual area and vice versa for the right half (see Fig. 3.3).

From LGN visual information passes to the visual areas of the cortex at the back of the brain. Here further highly complex processing takes place. Anatomists have identified different parts of the visual cortex and named the visual areas V1–V5. Several models of processing in the different areas have been proposed, but at this time there is no consensus. Lesions in area V4 of the cortex result in achromatopsia, complete color blindness. If the retinas and area V1 are intact the patient continues to maintain knowledge of form and motion. On the other hand, the patient typically experiences colors but not form, depth, and motion if, for example, as a result of nonfatal carbon monoxide poisoning, area V4 largely remained intact while other areas were damaged.

Visual maps are known to exist in other areas of the cortex, such as a region on the sides of the head in front of the ears. As many as 1000 different kinds of cells are located there, each responding to certain pictorial elements. When a certain group of these cells responds to stimuli we may recognize, as a result, a certain face or object. Most of the processing going on in the cortex is at the subconscious level, as mentioned in Chapter 2. There is a growing amount of evidence that the visual system at the highest level may consist of two mostly independent systems. One of these, termed on-line (dorsal), is concerned with eye movement and maintaining balance based on visual input, as well as reacting to immediate threats or opportunities. The other, off-line (ventral) system, is concerned with processes requiring more time, such as the conscious recognition of objects one has become aware of. Such findings point to elaborate systems that developed as a result of evolutionary processes and that how allow us to react appropriately in many possible different situations, such as playing tennis, attending a party, or contemplating art in a museum. Rensink has proposed a taxonomy of visual processing (Fig. 3.4) that describes different kinds of vision and how they relate (3). It provides an indication of the complexity of the visual process and its interaction with the operation of the body and consciousness. Color vision is a part of perception and is processed both unconsciously and consciously, but probably only off-line.

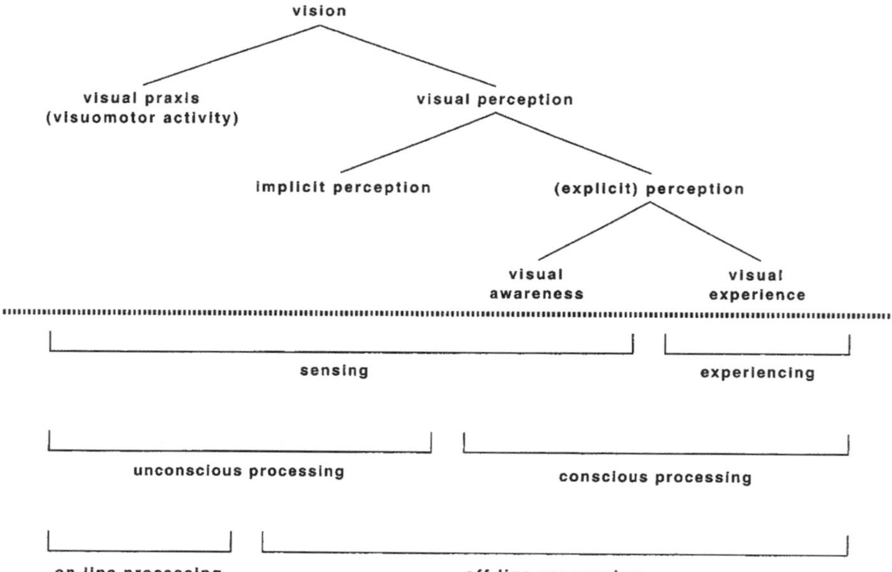

FIGURE 3.4 Taxonomical chart of human visual operations. The upper portion shows the relationship of the various operations, the lower portion their involvement in different kinds of processing (3).

The large amount of detail that has become known about the architecture and functionality of the color vision system may be only a small fraction of what needs to be known for a full understanding of its operation. New tools, such as scanning of functioning brains during execution of certain tasks, provide previously unavailable details. But as discussed in Chapter 2, the ultimate elaboration of color in consciousness remains a mystery.

RODS AND CONES

The active chemical substance responsible for transforming light quanta absorbed by rods or cones is well known and called *retinal*. It is a dye with a purple appearance attached to different kinds of protein molecules to form the four visual pigments of the normal human visual system. The proteins, among other things, fine-tune the spectral absorption properties of retinal. The visual pigment in rods is called *rhodopsin* (4). When absorbing light quanta retinal undergoes a molecular change making it colorless and, in a complicated sequence of events, triggering an electrochemical response in the rod receptor. This response is passed on to subsequent cells in the manner loosely sketched earlier.

The likelihood of a photon being absorbed by retinal in a rhodopsin molecule depends on its energy level. The greatest chance exists if the energy level corresponds to a wavelength between 500 and 510 nm. At higher or lower energy levels, the chances of absorption are reduced. The curve in Figure 3.5 illustrates the absorption characteristics of rhodopsin. At an energy level corresponding to 600 nm a photon has a ten times reduced chance of being absorbed compared to 500 nm. If it takes one photon at 500 nm to cause a given rod receptor to respond, at 600 nm it will take on average ten photons to cause the receptor to respond.

The figure also contains results of perceptual tests. The persons tested have been dark-adapted, that is, for an hour or so they sat in the dark before beginning the test. The test consisted of determining at different wavelengths the number of photons required to detect a light flash in an otherwise dark field of vision. The results of the experiment match closely the measured absorption curve of rhodopsin, except for the short wavelength area, and reveal the dependence of a fundamental visual response on the characteristics of the photosensitive absorbing substance (rhodopsin). Accurate determination of the response properties of rhodopsin is aided by the fact that at the very low light intensities used in the test, cones are not active because they have a higher response threshold.

Determination of the absorption characteristics of the active chemicals in the cones is much more difficult because of the overlapping response profiles as well as the fact that their number is much smaller than that of rods, and that they are irregularly distributed in the retina. The absorbance curves have been measured in primates as well as reconstructed from perceptual tests of humans corrected for the average absorption properties of the media in the eye from the cornea to the retina. The result is shown in Figure 3.6. The chemicals are sometimes given Greek-derived names: cyanolabe for the S pigment, chlorolabe for the M, and erythrolabe for the L pigment.

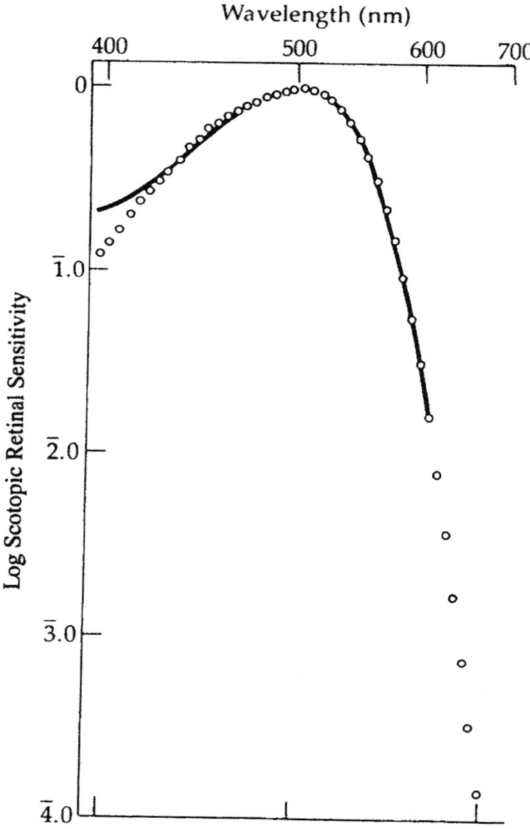

FIGURE 3.5 *Absorption characteristics of rhodopsin (solid line) and relative sensitivity of the dark-adapted human eye (circles) (5).*

The implicit cone sensitivity functions were measured for the first time with good accuracy in the laboratory of the German physiologist Helmholtz in 1886. Since then, increasingly refined measurements and calculations have been made to arrive at the data of Figure 3.7. The curves are shown on a normalized basis in a linear sensitivity scale; in all cases the area under the curve is the same. Cone sensitivity functions show the likelihood of photons of any given wavelength being absorbed. S-cone sensitivity is comparatively narrow and limited to the shortwave range. M- and L-cone sensitivities are much broader and overlap to a significant extent. Light of wavelengths between 400 and approximately 560 nm is absorbed to a larger or smaller extent by all three cone types. Above 560 nm absorption is essentially limited to two cone types only. Light at any single wavelength can be defined in the form of three numbers representing absorption in the three cone types. Lights with energy across the whole spectrum (such as daylight) or broad ranges of it can be expressed in terms of the same three numbers, where each number, in a normalized manner, represents the total cone absorption. Cone sensitivity functions can be seen as filters

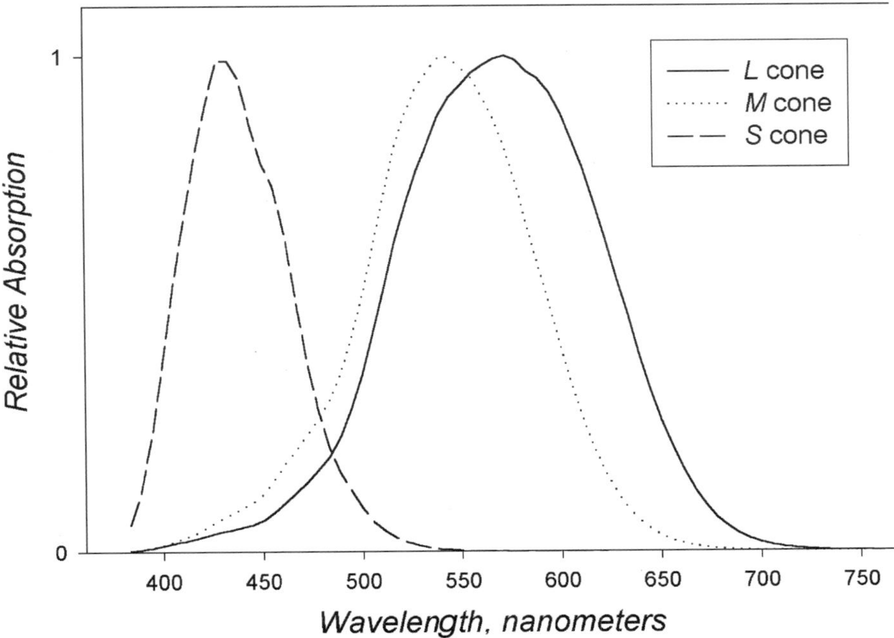

FIGURE 3.6 Average absorption characteristics of the three cone types in the human retina. Data from Color Vision Research Laboratory Web site (www.cvrl.ioo.ucl.ac.uk).

that take spectral power functions of lesser or greater complexity (single wavelength to full spectrum) and reduce that complexity to three values that (with an important limitation due to metamerism; see Chapter 6) are specific for every one of the spectral power functions (6).

The three cone types are represented in the foveal region of the retina in an approximate average ratio of $L{:}M{:}S = 32{:}16{:}1$, but can very significantly vary in individuals (7). Interestingly, this variation appears to have little or no effect on the color vision properties of the persons involved. This is an indication that normalizations take place at later stages.

It is important to keep in mind that the information generated at the cone level is immediately modified in the various cell types of the retinal layer described earlier. The three cone absorption numbers, therefore, have little direct relevance for the color experience we ultimately have when looking at a certain patch. They are like an important cooking ingredient (say an egg) changed beyond recognition in the making of a soufflé. Because the cone sensitivity functions have been quantitatively determined, and we can measure the energy levels arriving at the eye, together they are our only direct quantitative indicators defining the stimulus, the source of our color experiences, as "seen" by the average cones.

As a result, while it is possible to correlate, for both rods and cones, certain basic perceptual responses determined under controlled conditions quite closely with light stimuli, much additional behavior of the color vision system, as disclosed by

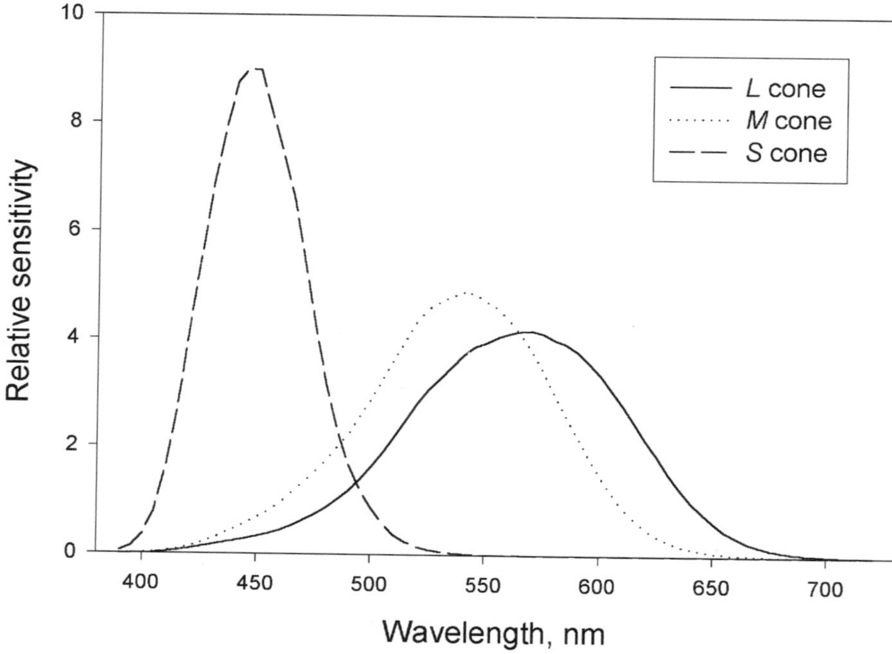

FIGURE 3.7 Average total spectral response functions of the three human cone types, derived from the Stiles and Burch 10° standard observer data. The functions are normalized to have equal areas under the curves. Data from Color Vision Research Laboratory Web site (www.cvrl.ioo.ucl.ac.uk).

perceptual testing, is not explainable simply on basis of receptor absorption characteristics. One issue relates to the fact that one of the basic attributes of visual perception—brightness or lightness—is not directly related to the signal from one of the cone types, but is now generally assumed to come from two. In a cursory way, the brightness signal is taken to be the relative sum of the signal output from L and half of the output from M cones. That this cannot be the whole story is indicated by the fact that, given the right circumstances, any "white," "gray," or "black" sample can be seen as having any of the appearances of white, gray, or black.

COLOR OPPONENCY

Based on measurements in individual LGN cells of macaque monkeys, vision scientists are now often expressing their data in a system where, in two dimensions, output data from the three cone types are subtracted from each other: in one case $L-M$, and in the other $L+M-S$. (As mentioned already, the third dimension, brightness, can be calculated as (on average) $L+0.5M$.) The subtractive dimensions represent a so-called *opponent system*. Opponent color processes were first postulated by the

German physiologist Hering based on his psychological insight of three pairs of opposing fundamental color perceptions: red and green, yellow and blue, and black and white. He defined unique hues as those that do not carry any trace of another hue. He found that there are only four of these in the hue circle. Unique blue is a hue that is neither reddish nor greenish, and comparably for the other three. For humans there is no direct relationship between the input into opponent-type cells in macaque LGN (believed to exist in humans also), mentioned earlier, and unique hues. In fact, individual choices of color stimuli representing unique hues vary to a surprisingly large extent. In addition, an opponent system for the black–white dimension has not been identified. Further, there is evidence that color-sensitive cells in the visual cortex no longer carry information in the same opponent form as LGN cells. On an empirical level, simple opponent color systems are a part of nearly all mathematical models of color vision. Chromatic color opponency is clearly a component of the human color vision system, if not in the simple form proposed by Hering.

In the second half of the twentieth century, perceptual opponent responses have been isolated using hue cancellation experiments (8). These experiments are based on the finding that lights that appear colored, when progressively mixed with complementary lights (appearance of opposing hue) result in progressive diminution of color, ending in its absence (white). For example, a light of unique blue hue is mixed with lights of various wavelengths, appearing more or less yellowish (lights of wavelength approximately 500 to 650 nm). The amount of blue light required at each wavelength to cancel the yellow component (until the combined light no longer has a yellowish or bluish appearance) is recorded and taken to be the a measure of the yellow response. Blue, green, and red responses were determined comparably. Unsurprisingly, the results vary by observer. Figure 3.8 illustrates the results for one

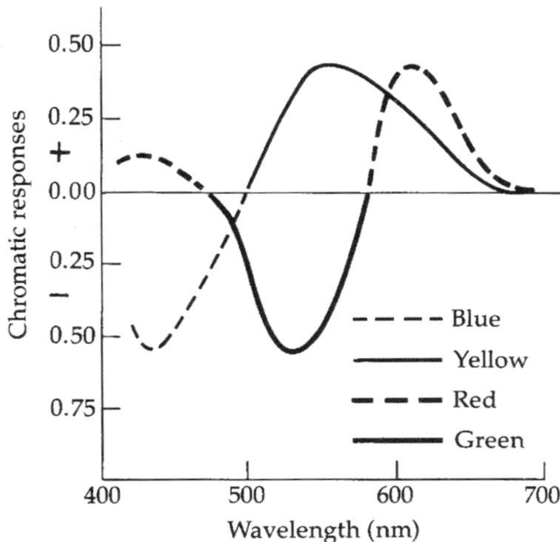

FIGURE 3.8 Results of cancellation experiments illustrating the imputed response of yellow, red, blue, and green opponent color mechanisms (8).

observer. Such functions can be modeled moderately well from cone responses, as Chapter 6 will show. Color opponency clearly plays a role in color vision. This role is most likely not exactly the one assigned to it in simple models such as the CIELAB color space and difference formula (Chapter 6), where calculation of the opponent responses is not based on scientific evidence for implied cell connections.

There is considerable and growing knowledge about the physiological mechanism of color vision (of which this chapter gives only the briefest of summary) (7). However, it is far from complete. There is growing evidence, however, of the lack of a simple relationship between cone responses and perceived color. As described in the next chapter, the visual system appears to be constructed to extract with good accuracy essential information contained in the spectral characteristics of the natural world. It uses its own (subconscious) interpretive mechanisms to come up with the most likely scenario of what the scene contains so that other mechanisms can use the information to plan actions or reactions. At the conscious level these capabilities leave room for much pleasurable experience.

4

Color Perception: Phenomena

Since the nineteenth century psychologists have been pointing out that our experiences of color fall into different categories. What these categories are has been under discussion since that time. Color science distinguishes between unrelated and related colors (to be discussed presently). The likelihood of two fundamental visual systems, one concerned with the moment-to-moment operation of the body and the other one available for contemplation, planning, and pleasure was mentioned in Chapter 3. In Western culture there is a tendency, since Aristotle, to regard color as a separate phenomenon that can be analyzed independently of conditions and tied to specific light stimuli. Many experiments have shown, however, that the color experiences we have are the result of the total situation in which they are obtained. The perception of given local stimuli varies if they are taken as induced by a light source or as due to a material. This fact gives rise to a theory that a major perceptual distinction by our visual system is in terms of coding of color as lights or as surfaces (1). The change in stimulus required to directly experience a just noticeable difference in brightness of lights is larger than the change required for a just noticeable difference in lightness of two objects. Rooms appear to shorten in depth when lit with intense chromatic light (2). There are countless more examples of this kind, showing that experiences resulting from stimuli are flexible and apparently elaborated by unconscious systems that analyze incoming stimuli in the context of the complete two-dimensional light array on the surface of the retina. It means that color is simply one interdependent aspect in an array that includes place, form, motion, and color. Much of color science of the last one hundred years has been concerned with establishing the relationship between isolated color stimuli and the resulting color experience. A future general theory of

Color: *An Introduction to Practice and Principles, Second Edition*, by Rolf G. Kuehni
ISBN 0471-66006-X Copyright © 2005 John Wiley & Sons, Inc.

color perception must be able to explain the results that have been established in this endeavor also, but it must do much more than that.

To be able to build mathematical models based on cone response, data that are accurate in a given specific situation, requires consideration of certain perceptual phenomena. Such phenomena disturb any simple linear relationship between stimulus and perception. A few of these have become apparent only upon finding regularities in perceptual data that could not be accounted for in simple linear relationships between stimulus and response. It is with this kind of color perception phenomena that this chapter is concerned.

UNRELATED AND RELATED COLORS

The most complete reductive change from natural viewing conditions is experienced in the so-called aperture mode, where colors are seen as unrelated. Practically, this is achieved by viewing a uniform color field through a narrow tube of black paper, by using a so-called reduction screen (a sheet of black construction paper with an opening cut in the center), or by viewing a color field on an otherwise black monitor in a dark room. Such fields have the appearance of lights, and the situation is similar to experiencing a colored light at night. The spectral colors seen by Newton in his experiment sketched in Figure 1.6 are (in a way) unrelated colors. Unrelated colors are rare experiences, mostly obtained in a laboratory. A stimulus experienced as brown as part of an object in natural surroundings is seen as orange or yellow in aperture mode. The transition between unrelated and related states is fluid. Stimuli reflected from an object can be seen as unrelated when viewed through a black tube. An unrelated stimulus (depending on its intensity) can be seen as an object color if it is viewed, say at night, through a white screen that is locally illuminated with white light. In bright daylight a neon light appears as an object indistinguishable from some others that are not internally lit. At dusk, as the surround light level decreases, the color perceptions from such low-intensity lamps become progressively brighter and more saturated. At a certain point they begin to look fluorescent and then glowing, taking on the appearance of lights. The three perceptual attributes conventionally applied to unrelated color percepts are brightness, hue, and saturation. Brightness is related in a complex fashion to the level of photon flow, hue to the relative content of given wavelengths in the total flow, and saturation is a measure of the degree of dilution of spectral lights with white light.

Of more practical importance are related colors, seen in the object mode. They are the colors we experience when looking at objects of any kind in more or less natural surroundings. Viewing conditions may be entirely natural such as we experience on an outing into the woods, mountains, or even a desert. Or they may be simplified to the extent of looking at two samples placed on a stand at 45° in a light booth illuminated with a standard light source. Here the conventional attributes are hue, blackness, and chromaticness, or hue, lightness, and chroma. The former relate to Hering's presumably more natural system where any color experience can be described, as mentioned, as the sum of fundamental chromatic, white, and black experiences. The latter are the subject of the following sections. An illustration of the relationship of hue, lightness, and chroma in the Munsell system is shown in Figure 5.8b.

LIGHTNESS

Lightness is technically defined as the perceived brightness of a nonwhite object compared to that of a perfect white object. Lightness perception is a very complex process from the viewpoint of our current knowledge about the operation of our vision system. Figure 4.1 is an illustration of one aspect of this complexity. The figure indicates that, depending on the implied context of form, illumination, and shadow to identical stimuli, vastly different perceptions can result. At the same time objects seen under more or less natural conditions approximately maintain their perceived lightness compared to that of other objects, regardless of the amount of light or the lightness of the surrounds they are seen in.

Technically, light is measured with a photometer, an instrument that measures the intensity of a stream of photons after it passes through a spectral filter that simulates average human brightness perception as determined by a particular method. This method was selected so that the measured brightness of two lights add up in conventional (linear) fashion when they are combined. Brightness measured in this way is called *luminance*. Its scale is open-ended because light intensity can assume very high values, much beyond what our eyes can tolerate, for example, direct sunlight. In the case of object colors, the comparable measure is luminous reflectance with a range limited to 100, the luminous reflectance of the ideal white object. As is evident from Figure 4.1, luminous reflectance can only have a reasonably straightforward relationship to perceived lightness under very simple conditions of viewing and surround.

(a)

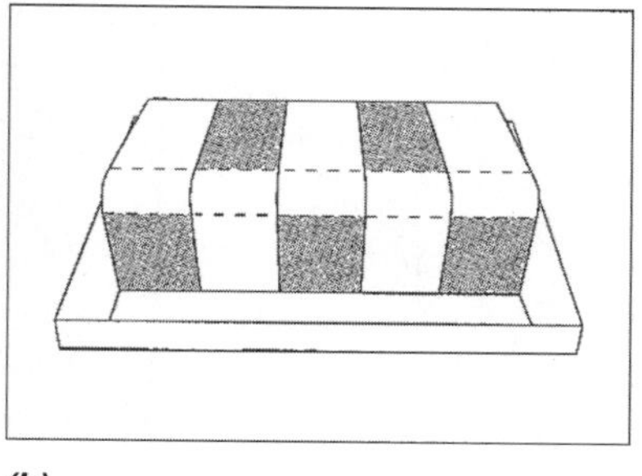

(b)

FIGURE 4.1 (a) Computer-generated image of a box on a table top, illuminated from above and behind, before a varied background. (b) Fields colored in gray on the accompanying sketch have identical reflectance properties (metric lightness $L^* = 54$*), but result in distinctly different appearance (3).* Figure also appears in color figure section.

For technical applications, such as color control in goods production, metric lightness is calculated from luminous reflectance (see Chapter 6).

Helmholtz–Kohlrausch Effect

At least since the mid-nineteenth century it has been noted that stimuli resulting in chromatic color perceptions require a lower luminance or luminous reflectance to appear to be equally bright or light than stimuli resulting in achromatic perceptions. This effect has become known as the Helmholtz–Kohlrausch effect (HKE), named for two important investigators of the effect. According to Helmholtz, chromatic colors have a glow of their own, independent of luminance or luminous reflectance. The degree of glow varies with hue and saturation of the perceived color. More recently, it has been discovered that the magnitude of the effect depends on how the experiment is conducted. This may be an indication that the magnitude is the outcome of a more general interpretation by the visual system of the complete context of the scene in which samples are represented. Figure 4.2 is an illustration of the HKE. The effect can be modeled with good accuracy by adding fractional amounts of opponent color functions (see Chapter 3) to spectral luminous reflectance. However, different experimental data require different amounts and components for best fit (4). The implication is that Hering's four fundamental chromatic colors, if they could be viewed at zero luminous reflectance, would not be black, but would be experienced as very dark but highly chromatic colors in a way that is difficult to imagine. To date HKE has usually not been considered in technical color models, such as color space and difference formulas.

Lightness Crispening Effect

The lightness crispening effect also has been known at least since the nineteenth century. It describes the fact that in order to see lightness differences of a given

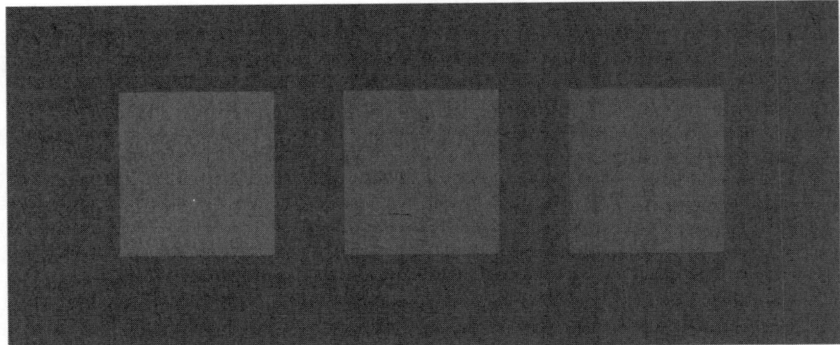

FIGURE 4.2 Examples of the Helmholtz–Kohlrausch effect. The three colored fields have the same metric lightness (CIELAB L) as the gray surround, but appear noticeably lighter.* Figure also appears in color figure section.

FIGURE 4.3 *Demonstration of the lightness crispening effect. The six gray fields have metric lightnesses of 20, 30, 40, 50, 60, and 70. The surround metric lightnesses are 65% on top and 35% on the bottom. The perceived appearance of and differences in magnitude between the fields varies as a function of the surround lightness.*

size (just noticeable differences, or Munsell-value step-size differences, for example; see Chapter 5), the smallest change in stimulus intensity is required if the luminous reflectance of the samples being compared straddles that of the surround in which they are viewed. If the luminous reflectance (or brightness for lights) of the surround is much higher or lower than that of the samples, the difference in their luminous reflectance has to increase in order to result in a perceived difference of equal magnitude. This is illustrated in Figure 4.3. As a result, a gray scale, a series of samples representing average perceptually equal distances between samples, can only be valid for a surround of specific lightness (5). Surprisingly, the metric lightness scale used in color difference formulas is a scale that does not apply to any specific surround condition, but rather to surrounds varying with the luminous reflectance of the samples (a situation never encountered in practice). So far, inaccuracy resulting from this situation is outweighed by the convenience of having a single scale. There is also a chromatic crispening effect to be discussed later.

HUE

Hue is defined in Merriam-Webster's Collegiate Dictionary (tenth edition) as an "attribute of colors that permits them to be classed as red, yellow, green, blue, or an intermediate between any contiguous pair of these colors." Hues are the most prominent aspect of a chromatic color experience. Their sequence is given by the spectrum. There are hues that do not exist in the spectrum but can be generated by mixing lights from both ends of the spectrum in different ratios. They are generally known as purples.

The totality of hues naturally arranges itself in a circle, since there is reddishness at both ends of the spectrum. The definition refers to Hering's four fundamental hues that in a psychological sense cannot be mixed from other colors. In terms of color stimuli this does not apply since, for example, a hue appearing as unique blue can be mixed from a reddish and a greenish blue stimulus (albeit with a loss in saturation). The same applies to a mixture of colorants (here the loss of saturation is likely to be higher, depending on the colorants used).

Hues can be scaled in different ways. The two most important are Hering's and according to equal perceived magnitude of hue difference between individual hues. In the former case, the four unique hues are placed on the axes forming a cross in the hue circle and the differences are scaled perceptually so that there are uniform increments/decrements of unique hue content in intermediate hues, such as in the Swedish Natural Color System (NCS; see Chapter 5). In the latter case, the perceptual difference between two neighboring hues is picked as a reference difference, and subsequent hues are picked so that the differences between them are all of the same perceptual magnitude as the reference difference. An approximate example of this is a Munsell hue circle at constant value and chroma (see Chapter 5). The two circles differ significantly. Samples picked as having unique hues do not fall on quadrant axes in the latter case and the number of standard-size hue differences between unique hue samples varies. The implication is that in NCS hue differences in all four quadrants are of different perceptual magnitude.

An interesting question is if people with color perception abilities considered normal have the same hue experiences when looking at a given stimulus (say, a Munsell chip). In several recent experiments surprisingly large differences have been found when subjects were asked to pick stimuli representing the unique hues for an individual (6). Unique hues are the only hues where such a comparison is possible, because there are no other universal references. The variability is particularly large for the unique green hue. In general, it is considerably larger than could be expected from variability among individuals in the placement in the spectrum of their cone sensitivity functions (see Chapter 3) and its causes are not known yet. Implications of these findings for models of color vision have not been fully considered yet.

CHROMA

The third conventional attribute of object colors is chroma, a specific case of the general attribute saturation. One definition of chroma is "attribute of color used to indicate the degree of departure of the color from the gray of the same lightness" (7). While hue represents the qualitative aspect of a chromatic color, chroma represents a quantitative aspect. This usage of the term chroma was introduced by Munsell and has become widely accepted. In the Munsell system the chroma scale occupies radial lines on a constant-value color chart. Recent experiments have shown that among the three attributes hue, lightness, and chroma the last is the most difficult to assess. In three large-scale experiments of the mid-twentieth century, circular contours of constant chroma based on thousands of individual judgments varied quite significantly (see

Figure 5.12). The reasons for these differences are unknown, but may involve the specific experimental techniques involved, or perhaps the observer panels used.

It is of interest to note that the relative change in cone activation needed to perceive a unit difference of chroma is twice as large as the change required for perceiving a unit hue difference of the same magnitude. This may indicate that hue and chroma differences are assessed by different systems in the visual cortex.

Chromatic Crispening

A phenomenon comparable to lightness crispening also applies to chroma. The change in cone activation required to perceive a unit chroma difference is smallest if the chromas of the two fields straddle that of the surround. If the surround is achromatic (white, gray, or black), the smallest change is required for a chroma difference to be seen between slightly tinted grays. As chroma increases against the surround, increasingly larger changes are required. This effect is evident in all experimental color difference data where small differences have been involved. However, unlike lightness crispening, known to apply at any unit size of difference, chromatic crispening fades as the size of the unit difference increases. It has faded at the magnitude of Munsell double chroma steps as the unit chroma difference (8). More on lightness and chromatic crispening in Chapter 6.

GRAYNESS

Object colors can be seen as having a gray content, as Hering has shown. The degree of perceived grayness also depends on the lightness of the surround.

This also applies to light sources. If the perceived brightness of a light is less than that of the surround, it is not seen as a light but that it contains grayness. As its brightness increases there is a point where the gray content is zero; above that it begins to look luminous. The point of zero grayness depends on hue (9).

ADDITIVE AND SUBTRACTIVE STIMULUS MIXTURE, COMPLEMENTARY COLORS

If two lights are mixed together, the result follows the rules of additive color mixture. The field showing the result has a luminance (not perceived brightness) that is the sum of the luminances of the two mixed lights, and it appears brighter than either of the individual lights. Its chromatic appearance is also different from that of either component light. Light of 520-nm wavelength (typically seen as greenish in appearance) and light of wavelength 650 nm (seen as reddish), when mixed together in an appropriate ratio, result in a yellowish appearance. Mixing light of 450 nm (reddish blue) with the same 520-nm light results in a bright turquoise appearance. When mixing light of 470 nm (blue) with light of 575 nm (yellow) in different ratios, beginning with mostly 470-nm light, the pure spectral color of blue begins to be desaturated and increasingly whitish. At a given ratio all chromatic appearance has disappeared and the resulting appearance is colorless. As the ratio tilts in favor of the 575-nm light, the appearance begins to be yellowish, ending in the pure spectral yellow of 575 nm. Pairs of stimuli (and their associated color experiences in a given situation of surround) of this kind are called *complementary*. For every spectral stimulus there is a complementary stimulus. Some of these are not found in the spectrum but can be mixed from lights at both ends of the spectrum. Complementary stimuli are located in the CIE chromaticity diagram (see Chapter 6) on lines passing through the white point of the diagram. Such stimuli have also been called *compensatory*. For most people their choices of unique hue stimuli are not compensatory.

Instead of mixing lights, one can mix colorants that affect the reflection properties of materials to which they are attached. This is known as *subtractive mixture*. Unlike lights, when mixing pigments the result is always darker than each of the components by itself (thus the name). When mixing a yellow, a red, and a blue appearing light in appropriate ratios, the result is colorless, or white when reflected from a white surface. When mixing yellow, red, and blue pigments in appropriate ratios and viewing the result in daylight, the appearance is dark gray or black. This situation had confused writers about color for over 2000 years until Helmholtz provided the explanation in the later nineteenth century. Comparable to complementary lights, painters have also called pigments complementary if their mixture in appropriate ratio results in a gray color appearance. Complementary pigment pairs can desaturate each other until chromatic color completely disappears, as the painter Runge described in 1810 and demonstrated in his color sphere (see Fig. 10.5 and Note 10).

ADAPTATION

The human visual system has the remarkable ability to adapt itself to a degree to the prevailing average quantity and quality of light. This complex process is known as *adaptation*. The result of adaptation is that despite considerable changes in the intensity and quality of the illuminating light, the effect on the perceived color of many objects is small or negligible. However, negligible effects are largely limited to natural lights and natural objects. It is under those conditions that the ability to adapt developed a long time ago. In the world of artificial light sources and artificial materials adaptation is much more limited.

If our visual system is properly adapted, various types of fruit in a glass bowl will be perceived as essentially looking the same in the bright sunlight of a clear day, in the diffuse light of an overcast day, or even in the light of a tungsten light bulb at night (all are approximations of blackbody radiations). Similarly, a piece of white paper will appear white under those conditions, and even in artificial white lights, despite large changes in the quantity and spectral quality of the lights.

Adaptation to quantity of light is easier to comprehend than adaptation to light quality. The metaphor (inexact) of a camera can be used to illustrate aspects of the former. In the case of a camera, the size of the opening (aperture, *f*-stop) behind the lens is used together with shutter speed to control the amount of light to which the film is exposed. In the case of the eye, the *f*-stop adjustment is duplicated by the expansion or contraction of the iris, resulting in a larger or smaller opening. In the case of the camera, further adjustment is possible by selecting a more- or less-sensitive film. In the case of the eye, retinal processes manage the conversion of light energy to electrochemical energy in a way that reduces the difference between lightest and darkest object to a scale of about one hundred to one, even though the ratio of light intensities may be as high as one million to one.

Adaptation to the quality of light, its perceived chromatic color, is a more complex process. There are many ways to demonstrate it. A simple one is as follows: After spending some time in a normally lit environment cover one eye with an opaque patch and, keeping the other eye open, spend three or four minutes in a room lit only with a standard red darkroom light bulb. Return to the normally lit room and observe that everything has assumed a greenish coloration. Close the eye, remove the patch from the other eye and observe that, seen with this eye, all objects have their normal perceived color. Observe further that after a few minutes the eye partially adapted to red stimulus light (it takes considerably longer than three or four minutes for complete adaptation) has returned to its normal state and there is no longer a difference in the adaptation state of the two eyes. Adaptation to red light has shifted the hues of all objects away from redness toward neutral to compensate for their light-induced reddish appearance. The direction of adjustment is opposite, that is toward greenness. This change is made visible by the sudden change to neutral illumination in which the objects temporarily assume a greenish coloration.

The example illustrates a fairly drastic case of chromatic adaptation. Adaptation is a continuous and ever-present process that normally goes unnoticed. Carefully designed experiments have shown that some aspects of adaptation take place very rapidly. Other aspects take a considerable amount of time. The longest time span is for dark adaptation, the condition where we can see faint stars in a dark sky or even a single burning candle at a distance of a mile or more. It takes an hour or more for complete dark adaptation.

The purpose of adaptation is not difficult to guess. It is of obvious use to a creature to be able to distinguish between friend and foe in the brightest sunlight as well as in murky twilight. It is also important to distinguish poisonous from edible fruit regardless of the time of day and spectral power distribution of daylight. The distribution changes sometimes relatively rapidly and at others slowly during the day, for example, when stepping from under the green canopy of woods into an open field, or when the sun slowly moves from the zenith toward the horizon. Continuous adaptation ensures essential color stability of perceived objects. When stepping into the modern world of artificially colored materials and artificial light sources the ancient adaptation system is often overwhelmed and we may experience more or less limited adaptation. This matter is discussed further in the section on color constancy.

The laws guiding adaptation have been and continue to be investigated. Different types of mathematical models based on cone sensitivity functions have been proposed in recent years and significant research activity continues (11). Adaptation cannot realistically be treated in isolation from the total perceptual process as current models do. Different adaptation models can perform about equally (modestly) well with given sets of perceptual data, an indication that the models are incomplete descriptions of the phenomena.

COLOR CONSTANCY

Color constancy is an important aspect of chromatic adaptation. Color constancy is experienced when, despite the distinctly yellowish coloration of the light from a common light bulb when first turned on at dusk, objects such as white paper or blue flowers maintain in essence their color appearance despite the drastic difference in the spectral signatures of the two lights. Objects are color-constant if their apparent color does not change (after allowing time for adaptation) regardless of the light in which they are viewed. There is, however, probably not an object in existence that can meet this very general definition. If a narrow band of spectral light or light filtered through certain kinds of filters is used as a light source, adaptation is incomplete and the apparent colors of objects will be different from those in daylight. Theatrical lights are well-known examples of such light sources. For color constancy to have meaning, the definition of light must be restricted to those sources similar to blackbody radiators.

The color experienced from viewing an object can be a function of the light it is viewed in. Color constancy (or the lack thereof) has been experienced by most people who have shopped for outerwear clothing. Two different garments that appeared to either match or were experienced as harmonic in the light of the store may no longer match or agreeably complement each other when seen in daylight.

While our color vision system may have been tuned a long time ago to maintain approximate constancy of the appearance of objects, it cannot do so now for all possible reflectance functions and light sources. This is an issue of considerable importance to artists, designers, dyers, graphic printers, and others. A painting created in the preferred natural north light of an artist's studio should ideally appear the same when bathed in the flood light in a museum. But this is only possible if the pigments and pigment combinations selected by the artist have reflectance functions that for the average observer result in very similar color experiences in the two different lights. To avoid surprises a painting to be viewed in incandescent light should be created in incandescent light. Similarly, garments that should look compatible in daylight should be purchased after viewing them in daylight. The problem of the effect of light sources on the apparent color of objects is well known to photographers. They work around the problem by either using a type of color film specially tuned to a given light source, or by lighting a scene (where possible) with a light source that "renders the colors of objects" in the desired way.

How to address color constancy in a technological manner has only become an issue in recent years because suitable mathematical models did not exist before. Color constancy is a special case of the more general problem of color appearance, and comprehensive color appearance models are of recent origin. They require experimental data showing how given reflectance functions are experienced by the average observer when the objects are viewed in different lights. From the results, conclusions can be drawn concerning what reflectance functions are required so that an object has a similar appearance in two different lights. The complexity of the problem increases if the object is to have the same appearance in several different lights, the desirable result. A mathematical model has been proposed very recently for such calculations (12). The success of this model is mixed, however. It is based on memory matching, that is, the subjects in the experiment view an object in one light, get adapted to another, and attempt to remember if the appearance is the same or how it has changed.

As mentioned color constancy is most evident in various phases of daylight. These are the conditions in which it developed by evolutionary process (see Chapter 2). The more the spectral signatures of light and objects diverge from natural lights and objects, the less color constancy obtains. It is not surprising that for some lights that sharply deviate in spectral composition from daylight, such as the sodium light of inexpensive street lighting or for objects colored with certain synthetic colorants, the degree of color constancy is much reduced.

Color constancy considers the problem from the point of view of objects. Conversely, the problem can be considered from the point of view of light sources. The applicable term in that case is "color rendering properties of light sources" (13). Light sources with certain blackbody spectral power distributions are considered standards and other light sources will render the perceived color of objects more or less different, and thereby have better or poorer color-rendering properties. A method for calculating a color-rendering index for a given light source has been developed.

Color inconstancy has been described already in classical Greece where the difference in appearance of purple-dyed fabrics in daylight and in the light of oil lamps was noted.

METAMERISM

One of the more surprising visual phenomena is related to color constancy, but involves at least two objects. It is applicable to both lights as well as objects. Two or more lights or materials are said to be metameric if they match (have the same apparent color) but have different spectral signatures (spectral power or reflectance functions). White-appearing light can be composed of rays of two wavelengths only or it can be composed of rays of the complete spectrum. A gray-appearing object can have a reflectance function that is a flat line across the spectrum or can have any number of much more complex ones, of which Figure 4.4 is one example. For the average observer in a given set of conditions materials with either function will appear to have the same color.

How is it possible for two objects to cause identical perceptions despite drastically different spectral signatures? The reason is found in the integrating properties of the three cone types. The normalized linear version of the cone sensitivity functions of Figure 3.7 all have the same area under the curve. If for each wavelength the reflectance functions of Figure 4.4 are multiplied with the corresponding cone function values and the results are summed, three numbers are obtained representing the effect of the radiation of a (for simplicity's sake) light with constant power across the spectrum

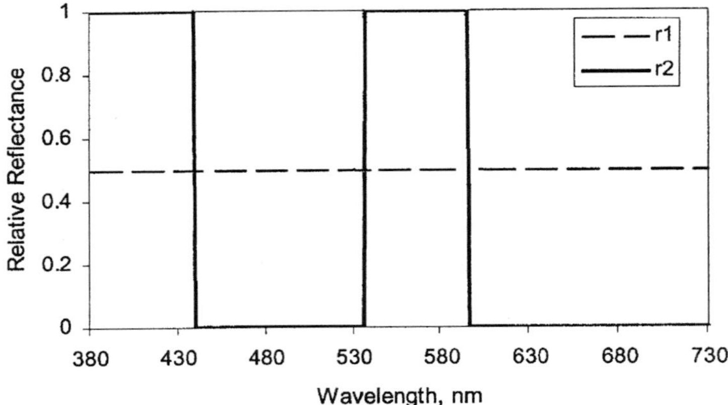

FIGURE 4.4 Two reflectance functions that are metameric for the CIE 10° standard observer. The (not shown) function where 0 and 1 values are reversed is also a metamer.

(an equal energy light) on the three cones. The numbers for both reflectance functions are identical. The appearance of the two objects matches. Equally, the appearance of a light of a single wavelength matches that of a light of broad spectral signature if the sets of numbers for the two lights agree. Metamerism, it turns out, is the automatic result of a process called *dimension reduction*: the reduction of the complexity of a broadband function (such as a spectral power function) to fewer dimensions, one or more of them, by filtering the complex function through filter(s) such as the cone functions. Regardless of the (limited) number and specific form of the filter functions there are always metamers: spectral functions of differing form that have identical values in the filtered format. However, the broadband functions that result in identical filtered values depend on the number and form of the filter(s). Spectral functions that are metamers for an organism with only one filter are not metamers for one with four filters. Similarly, spectral functions that are metamers for the average observer with the filters of Figure 3.7 are not metamers for a different observer whose placement of the three cone functions differs more or less.

For obvious reasons, in the case of objects, metamerism depends (in a complex fashion affected by the color constancy system) on the spectral power distribution of the light in which they are viewed. The triple of numbers is only identical for a given light source. If viewed in a new light source with a different spectral power distribution, the triple values are no longer identical and the appearance (most likely) is no longer the same. Such changes in appearance caused by light sources can be startling. They mean that one or both of the objects compared are color-inconstant.

Metamerism is a very fundamental aspect of color vision. It results in many technical problems. If a given color reference sample (say a chip from the *Munsell Book of Color*) is matched with pigments different from those used in the manufacture of the chip (a very likely situation), the resulting formulation is very likely metameric (unless an effort has been made to minimize it). Today, industrial color matchers are usually required to produce formulations that not only perform well in the application process and usage of the object but also match the reference in typically three different lights. Computer-assisted formulation is used to speed up this process. However, the reference sample may have poor color constancy, and as a result the matching formulations also will have poor constancy. In recent years the job of the color matcher has been complicated by requests not only for formulations matching in three lights but that are also color constant.

Metamerism not only causes the technical problems just sketched, but also makes possible technologies that have become very important to human society, such as color photography, color television (and monitors), reproduction of colored images with three or four colorants. These technologies are possible because of the filtering properties of human cones.

SIMULTANEOUS AND SUCCESSIVE CONTRAST, AFTERIMAGES

Experiments with a reduction screen, as described earlier, can give insight into how differently colored neighboring areas can affect each other's perceived color. Such effects are known as *contrast effects*, and they involve lightness as well as chromatic color. Lightness and chromatic crispening are kinds of contrast effects. Their evolutionary purpose was quite clearly to sharpen perception of neighboring fields with minor spectral differences that may signal the end of one object and the beginning of another (for example, the end of harmless greenery and the beginning of a greenish snake).

Contrast is fundamental to color perception. If all possible contrasts are removed from one's field of vision in a so-called complete Ganzfeld (see the Glossary), color perceived from a stimulus fades away after a few seconds. It is evident that with the exception of this very artificial condition, contrast is always present when our eyes are open.

In everyday life we are not aware of contrast because its effects are woven seamlessly into our fields of view, just as the effects of adaptation are or just as we are not normally aware of the blind spot. There is a time element to contrast effects, expressed in terms for the two basic kinds: simultaneous and successive. While in simultaneous contrast cause and effect happen at the same time, in successive contrast the present cause results in an effect several seconds later.

Figure 4.5 is an example of simultaneous contrast. Five identical squares are placed on a variable background (of variable chromaticity as well as lightness). Even though the so-called spectral return, the light returned to the eye, from each patch is identical, the apparent color varies quite dramatically. The perceived color assigned by our visual system to each patch is the result of the contrast between the cone responses to the light from the patch and the light from the surround. Simultaneous contrast moves apparent lightness and chromaticity in the direction opposite from those of the surrounding field. Of two gray adjoining fields of different lightness the lighter one will appear darker and the darker one lighter than if they are separated from each other. Their total color appearance is also affected by the lightness of the surround. Chromatic fields

FIGURE 4.5 *Example of simultaneous contrast. The five fields inserted into the varying surround are physically identical.* Figure also appears in color figure section.

change the appearance of an adjoining field in the direction of their complementary colors; thus, in Figure 4.5, the test patch against the reddish background assumes a greenish color and vice versa (14).

Understanding the effects of simultaneous contrast is obviously of considerable importance to designers and artists. Its effects have been studied and described, among others, by Goethe, and the French chemist Chevreul wrote an extensive text on the subject some 150 years ago (15).

An impressive example of simultaneous contrast is represented by colored shadows (16). It can be observed in nature or demonstrated as follows. An object is placed into the center of crossing flashlight beams, one without a filter and the other with a "green" filter in front. The combined light reflected from the white screen has a light greenish appearance, the shadow from the object in the path of the "white" light has a darker greenish appearance, while that in the path of the "green" light has a distinct reddish appearance.

Successive contrasts appear in the field of vision after exposure to a relatively strong stimulus. Successive contrasts can be achromatic or chromatic. A small window in a relatively dark room that appears bright white due to outside light when viewed for several seconds can result in a black image when the gaze is shifted away from it onto a white internal wall. Staring intensely for some twenty seconds or more on an image of an American flag in green, yellow, and black will result in an image of the flag with normal appearance if the gaze is shifted onto a white wall or paper. Such afterimages are called *negative*. The apparent colors are changed to their complementaries. Under certain conditions it is possible to observe positive afterimages, that is, images where the perceived color of the objects remains nearly unchanged. Afterimages can be experienced with open eyes, but in most cases equally well with closed (17).

SPREADING AND EDGE EFFECTS, MACH BANDS

There are several additional kinds of effects that appear in certain circumstances. Among these are spreading effects where color appears to leak out considerable distances from narrow colored bands. This effect can result in the perception of contours where none exist.

Sharpness of edges affects both perceived lightness and chromatic contrast effects. Sharp edges result in strong contrast, while fuzzy edges result in diminished contrasts. In images of adjoining uniform fields of varying achromatic or chromatic colors (known as the *Chevreul illusion*) the appearance of the fields is not uniform, because it is affected differently at each edge by the contrast against the particular adjoining field.

The tendency of the visual system to intensify edges in the field of vision can result in what are known as *Mach bands*. These bands appear distinctly even if the stimulus change at the point of the Mach band is minimal.

There are several Web sites on the Internet that show examples of these and many other effects. Among the more impressive ones are www.purveslab.net, www.michaelbach.de, and www.uni-mannheim.de/fakul/psycho/irtel/cvd.html. Others can be found with a Google search. The reader is encouraged to study these.

VOLUME COLORS, TRANSPARENCY, TRANSLUCENCY

Color stimuli can also come from transparent or translucent materials containing colorants, such as metal salts or dyes. A simple example of material displaying volume color is a glass of red wine. The glass is usually transparent and colorless, the wine contains natural dyes that give it a deep purplish color, by absorbing most of the photons in the middle range of the spectrum, say from 480 to 650 nm. Glass itself can be colored by the inclusion of metal salts during manufacture, as demonstrated in windows in Gothic cathedrals. Color transparency film is another material representative of volume color. Translucency, a state where some light passes through the object while the remainder is scattered on the surface or interior, is achieved by inclusion of pigments in transparent plastic or rough sanding of a surface of glass or plastic.

Volume colors have a unique appearance. Light passes through them and what is not absorbed can continue directly to our eyes. The degree of absorption depends on the concentration of colorant in the medium as well as the thickness of the layer of the medium. The variety of color perceptions is similar to that of object colors, but their appearance is different, because not limited to a surface. Special scales have been developed to assess the color of some liquids, such as the color of petroleum products by the Saybolt Chromometer, or the platinum–cobalt scale (18).

METALLIC COLORS

Polished metals are noted for a high degree of specular reflectance, like mirrors. Some metals, due to special processes briefly discussed in Chapter 1 absorb some of the light falling on them and appear colored. The combination of shiny appearance together with a few limited hues gives metals a characteristic appearance that sets them apart from other materials.

In this chapter several fundamental color vision phenomena have been briefly discussed. They fundamentally affect and vary the experiences we have from given physical stimuli. Interpretation of the effects is sometimes straightforward but usually complex, and causes are often as yet unknown. They are effects often not noticed in the everyday world but clearly noticeable when presented in extreme conditions. Many are likely effects resulting from the operation of higher-level brain functions with the task to interpret images of the world arriving at the retina in the most likely way based on past experiences of the species and the individual. The following chapter represents an overview of issues of placing all possible color experiences into orderly systems.

5

Orderly Arrangements of Color

Studies indicate that humans with normal color vision can distinguish among some two million different color sensations when viewed against a midgray background, and perhaps double that if the background is widely varied. The question of how to place these perceptions into an orderly and meaningful arrangement has been of interest for more than 2000 years. An opinion that proved influential until the seventeenth century was that of Aristotle. He believed colors to be generated from the interaction of darkness and light, and that there are seven simple colors out of which all others are obtained by mixture. The true meaning of the seven color names is not certain in all cases, and translations vary (see Chapter 10). They may have been: white (pure light), yellow, red, purple, green, blue, and black (pure darkness).

In the second half of the seventeenth century Isaac Newton demonstrated that a narrow beam of sunlight refracted with the help of a prism forms in a dark room a band of light that we experience, when reflected from a piece of white paper, as having many different colors. Newton, who was not only the preeminent scientist of his time but also an alchemist believing in universal harmony chose, in analogy to musical tones of an octave, to recognize seven hues in the spectrum: red, orange, yellow, green, blue, indigo, and violet (ROY G BIV, see Figure 12.2). But Newton's choice of seven was controversial for the next 200 years. Recent research has indicated that for people not professionally involved with colors the choice of seven different hues in a displayed full spectrum is average (1). Repeated tests in the last 150 years have shown about 120 discernible colors in the spectrum.

Spectral hues do not encompass all hues we can recognize. For about half of all people the unique red hue is not found in the spectrum but in a mixture of short and

Color: *An Introduction to Practice and Principles, Second Edition,* by Rolf G. Kuehni
ISBN 0471-66006-X Copyright © 2005 John Wiley & Sons, Inc.

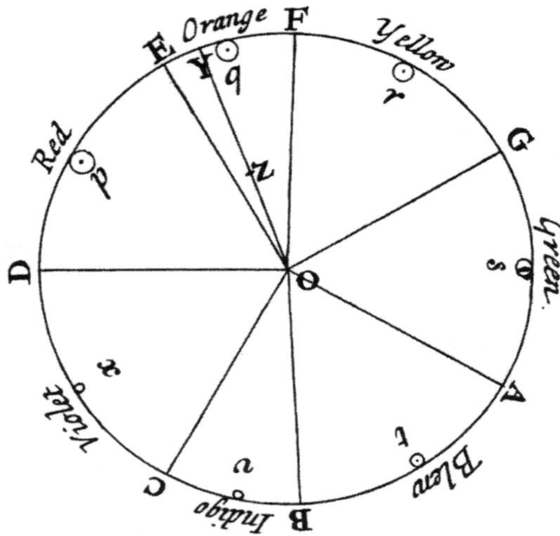

FIGURE 5.1 *Isaac Newton's color chart of 1704 with seven spectral colors arranged according to the chromatic tonal sequence. White is located in the center at O (2).*

long wavelength light. Such mixtures are seen as having hues from red to violet, depending on the relative amounts of the two kinds of light. In this way all hues fall naturally into a series returning upon itself: a circle formed by spectral hues and hues mixed from the spectral ends.

Already Newton recognized three aspects of color perceptions: brightness, hue, and intensity. Brightness describes how bright or dark a color appears, and saturation distinguishes spectral from desaturated and grayish color perceptions. In order to explain his results of mixture of spectral lights, Newton placed the spectral hues in a circular diagram with white as the common center and saturation lines as radial lines from the white center to the spectral periphery (Fig. 5.1). Painters and dyers already from before Newton's time believed in three fundamental chromatic colors that they sometimes equated with pigments or dyes: yellow, red, and blue, from which all other hues can be created. Experience taught that mixtures of three such colorants often are less saturated in color than those they were mixed from. The first illustrated hue circle based on pigment mixture was published in 1708 by an anonymous author in a book about miniature painting (see Fig. 11.2). Beginning with Mayer and Lambert in the eighteenth century the three-dimensional nature of color perceptions began to be illustrated systematically. They quickly learned that colorations obtained with given pigments do not follow the perceptual attributes in a simple way, making systematic coloring difficult.

As discussed before, in the later nineteenth century Hering reported on his introspective psychological evaluations of color and stated that there are six fundamental or unique colors (*Urfarben*): white, black, yellow, red, blue, and green. In none of these can a trace of any of the others be perceived. The unique red hue, therefore, is a red hue that is neither tinged with yellow nor with blue, and comparably for the others. In an orange hue we perceive redness and yellowness, in a gray, whiteness and blackness. In some color perceptions we can recognize up to four of the unique colors, for example, in a brown we may recognize yellowness, redness, whiteness, and blackness. Hering also stated that two of the fundamental colors always form opposing pairs, that is, white and black, yellow and blue, and red and green. He concluded this from the fact that one can recognize, say, redness and blueness in violet, or yellowness and greenness in chartreuse. But there are no hues in which he could recognize greenness as well as redness, or yellowness as well as blueness. Thus they are opposing pairs. (Opposing pairs of the Hering kind should not be confused with complementary colors discussed in Chapter 4.) The opposing pair of black and white is an exception, because we recognize both of them in grays. The ratio between blackness plus whiteness content in a color perception and the full color content is a measure of saturation. Newton's hue circle was not complete in Hering's sense because it did not place opposing hues opposite and it did not recognize a group of hues not found in the spectrum, the purples.

ORDERING COLOR PERCEPTIONS

To gain a better understanding of some of the issues in ordering object color perceptions it may be useful to go through a practical example. Picture yourself as an apprentice to a mosaics master. There are hundreds of differently colored mosaic pieces and you have been asked to place them in drawers in a sensible order. You could give each sort a name and place them alphabetically by name. But the master would have to learn hundreds of names to find those representing the colors he wants. The systematic ordering might instead be in terms of perceptual attributes. The most striking aspect of chromatic color perceptions is hue. You might begin with unique yellow and red. Reddish yellows, oranges, and yellowish reds fit in between. Arriving at unique red you realize that there are bluish reds and you continue through purples, violets, and reddish blues until arriving at blue. In a similar fashion you proceed via turquoise through green until chartreuse colors change into yellow. The question arises how many divisions to make around the hue circle and how to make the divisions. One kind of division is by equal perceived distance between grades, another (if you are familiar with Hering's unique hues idea) is by equal judged percentage change in the two unique hues involved. If you have used both methods accurately, you will find the results differ. A third method could be sorting complementary hues into neighboring bins, again giving different results.

In each hue category there are now a considerable number of pieces that appear to have the same hue but vary in lightness and at the same time are of more or less intense color (different in saturation). There is also a series of samples without hue, grays from white to black. These can be ordered according to how light or dark the gray is and result in a lightness scale. Assigning chromatic samples to lightness grades is somewhat difficult and you and the master may not agree on the result (with him prevailing!). The mosaic pieces now have been sorted according to hue and lightness, but you still have pieces of different apparent color in each category. You recognize that for each hue and level of lightness there are colors of smaller or greater chromatic intensity, starting with zero for grays and ending with the highest grade for the most intensely colored samples. You decide to name the third sorting attribute, the one for chromatic strength, chroma (as Munsell did).

One result of this sorting scheme is that the sample producing the most intense yellow color perception is found to have a different level of lightness (and likely also chroma) than the sample resulting in the most intense blue experience. The master may prefer to have all samples with highest chroma located on the same level. As a result lightness and chroma can no longer be attributes but, using Hering's approach, the samples can be sorted by full color and according to their perceived content of blackness and whiteness. In the former approach all samples can be placed into a cubic drawer system (perhaps similar to movable library stacks) in which hue forms one dimension, say width, lightness the second (height), and chroma the third (depth; Fig. 5.2). If a hue circle is preferred, the result is a cylindrical system with hue changing around the circumference, with gray in the center, lightness along the vertical axis of the cylinder, and chroma along radial lines (this and the following arrangements may be more satisfying but difficult to implement with drawers for mosaic pieces; Fig. 5.3). The Hering system, on the other hand, fits into a double-cone arrangement, with the most intense (full) colors on the periphery of the central plane, and the colors where whiteness exceeds blackness in the upper cone and those where blackness predominates in the lower cone (Fig. 5.4). Instead of Hering's double cone, there could also be a sphere (such as Runge's of 1810; see Fig. 10.5).

The experiment demonstrates that colors can be systematically arranged in different ways, each having a limited number of meaningful attributes. It appears that at least three attributes (and three dimensions) are required to systematically place all color perceptions. Attributes and dimensions imply scales, and the question arises what the scales represent. In our hue/lightness/chroma cube, the lightness scale can meaningfully represent equal perceived distances between lightness grades; the same applies to the chroma dimension and its scale. A lightness step can be made perceptually equal with a chroma step. But with hue we encounter a problem. Planes of colors of constant hue (but varying lightness and chroma) all have a common gray scale. The hue differences between colors of two neighboring planes change as a function of chroma: the perceptual distance between two reds of neighboring hues (for example)

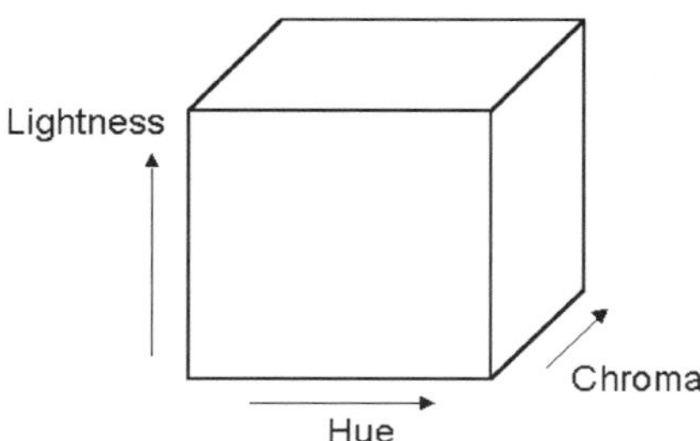

FIGURE 5.2 Cubic arrangement of object color perceptions according to the attributes hue, chroma, and lightness.

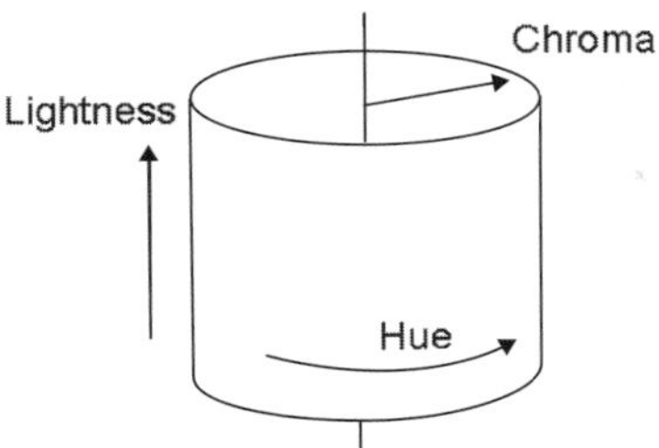

FIGURE 5.3 *Cylindrical arrangement of object color perceptions.*

is much smaller for two low chroma reds than for high chroma reds. As a result it is impossible to have a color cube perceptually uniform along its three axes. But, you say, if we use the cylindrical version with the common gray the problem is solved.

When Munsell developed his cylindrical system based on the attributes hue, lightness, and chroma early in the twentieth century he was interested in perceptual uniformity, but for practical reasons (a value [lightness] scale of 10, a hue scale of 100 with only 40 samples colored, and an open-ended chroma scale) the perceptual magnitude of the three scales differed. The Munsell system, to the extent that it is perceptually uniform, is only separately uniform in terms of hue at a given level of chroma and in terms of chroma throughout the system. The lightness scale of the modern system is uniform in terms of (compressed) luminous reflectance but not perceived lightness, as it does not consider the Helmholtz-Kohlrausch effect discussed in the previous chapter.

As a brief aside, it is useful to discuss the reason for the open-ended chroma scale. Hering had placed all his full colors on a common planar circle, that is, he took all of them to have the same chromatic power. Munsell, using a uniform chroma scale for all hues, soon learned that his pigments had different levels of chromatic power in terms of perceptually uniform chroma steps. Different hues ended up having different maximal chroma in the system. The same applies to spectral lights: their chromatic power differs by hue (3).

There is another important fact pertinent in the discussion of color order. As mentioned, scales imply differences and a perceptually uniform arrangement, if it can be

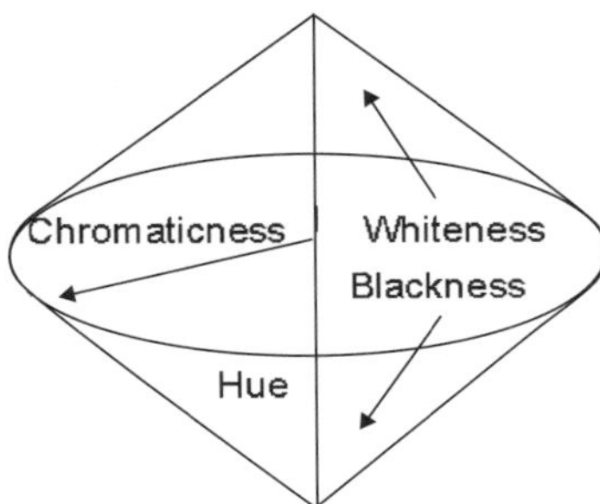

FIGURE 5.4 *Double-cone arrangement resulting from placement of colors of constant hue in triangles with a common gray scale as the central axis.*

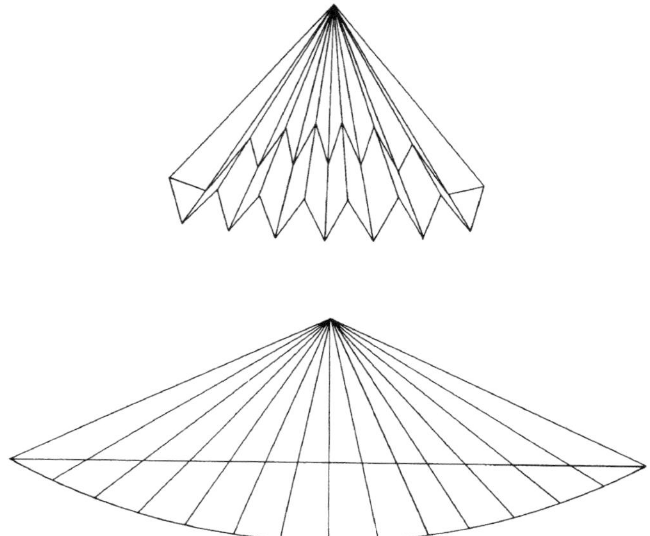

FIGURE 5.5 Judd's drawing of the crinkled fan demonstrating the effect of hue superimportance (5). On top the partially folded fan of a section of a hue circle, on the bottom the flatted version extending over twice the area.

captured in a mathematical formula underlying the geometrical model, implies the possibility of expressing the magnitude of perceived differences with numbers. In 1936 Nickerson developed the first color difference formula for the purpose of objectively expressing the degree of fading of colored textiles exposed to light (4). She used the Munsell system to express distances in hue, chroma, and lightness, but arranged the scales so that hue differences at a given level of chroma, chroma, and lightness differences were of comparable magnitude, and summed the three component differences. A few years later Judd investigated the geometrical structure behind the formula and found, to his surprise, that the implied total angle for all hue differences around the hue circle did not amount to 360°, but to slightly more than twice that number (5). In other words, if the unit hue difference at, say, chroma 5 perceptually equals the unit chroma difference, the system cannot be expressed in three dimensions. Judd named the effect "hue superimportance" and demonstrated its effect in the constant lightness plane as a crinkled circular fan (Fig. 5.5). Only a section of it can be flattened without overlap, but not the total fan. This turns out to be a key finding in the search for a uniform color space and will be discussed further below.

LEVELS OF COLOR ORDER

Psychophysicists, people interested in developing relationships between physical stimuli and resulting perceptual results, have developed a hierarchy of levels with increasingly higher levels of organization and information content (6). The lowest level is called *nominal* and provides the coarsest degree of organization. Many different kinds of perception of greenish colors are summed under the category name of green. At the next, ordinal level a qualitative ordering of all the greens is performed by, say, hue, chroma, and lightness without paying attention to quantitative scales within the attributes. Increasingly bluer greens are found in one direction, yellower in the other, lighter ones higher up and darker ones further down. At the next, interval, level quantitative attribute scaling is introduced. Until the Optical Society of America Uniform Color Scales system (OSA-UCS) was developed (see below) only intra-attribute uniformity was attempted. The next level in general scaling theory is magnitude scaling. In the Hering system judgments of the magnitude of full colors and of whiteness and/or blackness in a mixed color are made.

Here it is useful to talk about the meaning of intervals and magnitudes in regard to color. As mentioned, historically two sets of attributes have been developed that, for simplification, can be identified with Hering and with Munsell. In the former case full color content, whiteness, and blackness are considered magnitudes that are judged absolutely. These magnitudes are then divided into equal intervals. Perceptual distances are not a concern. The geometrical form of the system is decided in advance after arbitrary decisions have been made: distances between unique hues are equal and chromatic content of all full colors is equal. The solid is divided into portions of the basic magnitudes. In the case of the Munsell system, hue, lightness, and chroma are judged in terms of intervals only, and the system is built by adding up intervals from the central middle gray.

At the ordinal level orderly arrangements of color perceptions can take just about any desired three-dimensional geometrical form. Psychologists, artists, and color theoreticians of the eighteenth to twentieth centuries have produced a considerable variety of forms, such as the already encountered cube, sphere, and double cone, as well as double pyramids, tilted cubes, double cones, and spindle shapes (see Fig. 10.8). As ordinal representations they are all equally valid, but their interval organization varies widely, with usually no defined perceptual meaning. For the purposes of color difference evaluation and objective color control, the perceptually uniform organization contains more useful information than any other.

The most ambitious attempt to establish a globally uniform color space (perceptually uniform in as many directions as geometry allows) was made by the OSA-UCS committee from 1947 to 1975. As will be presented later, this system has a crystalline internal structure defining twelve equally distant points from any interior point and thus in principle assuring perceptual uniformity in these directions throughout its space. As the next section will show, the committee only partially succeeded in its endeavor.

UNIFORM UNIT CONTOURS IN EUCLIDEAN COLOR SPACE

An important question is what shape contours of uniform perceptual distance around a given point have in a Euclidean space. Perceptual uniformity makes this contour a sphere centered on the reference point. The implication is that all colors located on the sphere surface have perceptually equal distances from the center color. Many kinds of experimental evidence indicate that due to hue superimportance the unit contour is elongated in the direction of constant hue lines. This applies to all sizes of color difference, from very small to large. The exact shape is not known with certainty, but is usually taken to be an ellipse in a plane and an ellipsoid in space. Elongated shapes also have been established at the level of color-matching error, a kind of unit color difference smaller than we can perceive. It is obtained by repeatedly measuring the error in different directions in color space when attempting to match a given color represented by a point in the space (see also Chapter 7). In small color differences, such as are of interest in color quality control, most experimental unit contours, when represented in a Euclidean space, are elongated approximately in constant hue direction. Differences in the Munsell system are on average about five times larger than typical small color differences. Here unit difference contours implied by the Nickerson color difference formula can be plotted into the Munsell chromatic diagram, assuming they have elliptical shape (Fig. 5.6). In the OSA-UCS space, where the steps are approximately 1.5 times the size of the Munsell steps, the unit contours

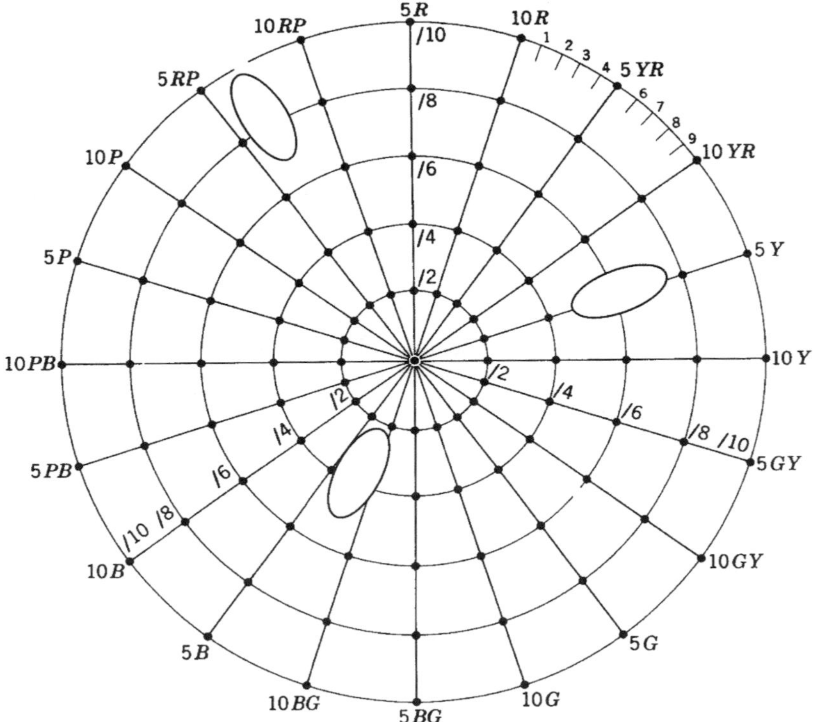

FIGURE 5.6 *Munsell perceptual constant value chart with three unit chromatic difference ellipses of arbitrary size.*

are by design of crystalline shape where the unit crystal fits into a sphere. This is not in agreement with all other color difference data. However, when investigating the experimental data on which the system is based, it is evident that unit contours should also here be elongated. After realizing the impossibility of fitting a system uniform in twelve directions into a Euclidean space, the committee decided to build instead the best approximation to a uniform Euclidean space, OSA-UCS (7).

Unit contours vary in elongation somewhat as a function of size of difference: the smaller the difference, the more elongated the unit contour. The average ratio between the major and minor ellipse diameter is approximately 2:1 (8).

The immediate reason for the elongation is hue superimportance, mentioned earlier. But what is its ultimate source? As yet the reason is not known. We can speculate that evolutionarily the detection of hue differences was more important than the detection of saturation differences. Our visual system requires less of a change in activation of the three cone types to see a difference when hue is involved than in the case of saturation. This points to separate subsystems for hue and saturation. It is known that lightness difference evaluation is served by another separate subsystem.

It is evident that in terms of a uniform or isotropic color space Euclidean geometry is not natural, but a cultural construct. Nonuniform color spaces can have many different geometrical forms, as was shown earlier. The fact that uniform color space is not Euclidean results in a degree of difficulty in the calculation of color differences, as will be shown in Chapter 7.

IMPACT OF CRISPENING EFFECTS ON COLOR DIFFERENCE PERCEPTION

Crispening effects were presented in Chapter 4. According to Schönfelder's law, and as mentioned earlier, differences between two colored fields are detected best if the surround of the two fields falls in lightness as well as chromaticness between the two test fields (9). This is important for the description of unit color differences and the calculation of the size of difference. In industrial practice, color differences are visually evaluated in light booths that usually have a light gray interior. This color is the basis for the crispening effects applicable under those conditions. Schönfelder's law means that for a unit perceived difference in hue, chroma, and lightness, the change in cone activation or the change in tristimulus values (see Chapter 6) is smallest for colors at or near the surround color (in our example, light grays or grayish chromatic colors). The required change increases in size as the sample colors differ more and more in any direction from the surround color. As a result, unit difference ellipsoids increase in size for colors more and more distant from the surround color. In practice this means a color space or difference formula is only applicable with good accuracy for the surround against which the perceptual data on which it has been based have been determined. For small color differences, the unit contour size increases by about a factor of 4 from near grays to the most highly saturated pigment colorations (10).

In Chapter 4 it was stated that the chromatic crispening effect is dependent on the size of the unit difference involved, while lightness crispening appears not to be.

Chromatic crispening is active at the level of color-matching error and at small to medium color differences. At the level of Munsell- or OSA-UCS-sized differences it has faded.

In the same chapter, the effect of lighting and surround conditions on the apparent color of objects was briefly discussed. In that sense the samples of the Munsell or any other color-order system have variable perceived color. They can be used as an accurate visual reference only under tightly controlled conditions of illumination and surround. The differences involved can be very large (as, for example, illustrated in Fig. 4.5), but we tend to disregard them in daily life.

OBSERVER VARIABILITY

As already discussed in Chapter 4, there is a considerable variation in color perception of objects by individual observers. A color chip seen as having a unique green hue by one observer is apparently seen as having a distinctly bluish green or distinctly yellowish green hue by other, color-normal observers, and comparably for the other unique hues. Experimental data indicate that unique hues of a given observer are for other observers not rotated equally in one direction or the other. The implication is (as yet not experimentally investigated) that perceptual distances in terms of the number of equal difference steps between unique hues can vary by observer. If found to be true, this will have a significant effect on individual color difference perception.

Psychophysical testing also indicates that individuals cannot operate as neutral color-measuring instruments, but each conscious decision regarding the perceived magnitude of difference between two differently colored fields (or any other) passes through some as yet unknown kinds of individual mental filters before judgment is rendered. It is likely that these filters represent conscious and unconscious personal intentions, biases, and experiences. For this and other reasons, color difference judgments can vary significantly by observer, and color spaces and related color chip collections as well as color difference formulas are based on an average observer. This usually means that test color differences are assessed by dozens of observers at least once, rarely multiple times. It is unlikely that the observer panel used for one color difference experiment is truly comparable to that in another experiment. In addition, there almost always are differences in samples, surround, and experimental method. Given these factors, it is not surprising that the results of different experiments in color scaling and difference assessment are not in good agreement, as will be seen in Chapter 7.

Sample Collections Representing Color Spaces

Historically, there have been and are many different kinds of color spaces and related sample collections. Conceptually, they can be sorted into three classes:

- Spaces and corresponding samples representing equal intervals of color stimuli of some kind.

- Spaces and samples representing equal intervals of response (perceived color).
- Spaces and samples representing some other concept.

A color stimulus is normally a particular spectral power distribution of light. In the case of related colors, it is a function of the reflectance properties of the object and the spectral power distribution of the light source. To produce any kind of broadband spectral power distribution of light on demand requires complex equipment. On the other hand, the reflectance characteristics of objects can be adjusted relatively easily (within limits) by using different colorants and varying their concentrations. However, as mentioned, colorants mixed in simple ratios do not follow perceptual attributes (hue, lightness, chroma) in a simple relationship (see Chapter 8). Achieving a series of colored chips that vary in agreement with attribute scales required in the past extensive trial-and-error work and today the support of computerized colorant formulation (see Chapter 9).

Color Space Sampling with Equal Intervals of Stimulus

An early tool of creating with relative ease many different color stimuli was an apparatus used by many investigators since its invention in the eighteenth century, employing Maxwell's disks, named for its most important user. As we know from movies and color television, the eye cannot resolve details of rapidly changing images, but creates time- and size-weighted averages. When a disk containing sectors of different colors is rapidly spun the visual system can no longer detect the borders of the sectors and the colors of all sectors are averaged according to sector size. By changing colors and sector sizes many stimuli changing systematically and regularly can be generated with relative ease. A problem is that the resulting stimuli are dependent on the colorations of the disk sectors and limited in chroma to the chromas of the corresponding colorants. Simple systematic changes in disk sector values do not result in perceptually equidistant steps. To relate the results to some absolute system of color specification requires elaborate calibrations and measurements. Maxwell disk apparatus have been used in the development of the Munsell and other color-order systems, where the perceptually selected disk-generated colors have been matched with pigments so that the resulting color chips have the same appearance.

A similar effect is achieved by presenting arrays of small colored lights where each light is so small in size that it cannot be individually resolved in the array. As a result, the visual system averages the color over regions limited by contrasting lines. Practical examples are color television and general monitor screens. By varying the activation of the three monitor phosphors or LED's, systematic series of color stimuli can be generated. The degree of resolution and the stability of monitors now available have resulted in their increased use in color vision research.

Other techniques with less reliable results involve half-tone printing. Several systematic samplings of colors achievable with particular printing pigments, for example, the standard printing pigments cyan, magenta, and yellow, were produced in the nineteenth and twentieth centuries, as mentioned earlier.

Color-Space Samplings with Equal Interval of Response

Because of the primacy of perceptions in the experience of color, the desire for samplings of color space according to perceptual distance is obvious. Perceptual distance can be along specific attributes, such as hue, lightness, and chroma, or blackness and whiteness. It can be uniform according to these or other attributes, but without the distances being of equal perceptual size in all directions. We can conceive of scales where only hue changes, while lightness and chroma remain the same, and similarly for the other two attributes. As mentioned earlier, such scaling does not result in a space uniform in all directions. Alternately, we can think of scaling color differences so that the perceived difference in any direction from the reference point is the same, a truly uniform color space. The term "uniform (isotropic) color space" should be reserved for this case.

There are several different methods of scaling and they usually produce somewhat different results. For example, a gray scale can be established by continually halving the perceptual distance between grades, starting with black and white. Another method consists in determining the number of just perceptual distances between white and black by measuring what change in luminous reflectance it takes for the first just perceptual step away from white and continuing in this fashion toward black. The relationship between luminous reflectance and perceptual scale in the two cases will likely be different. As mentioned, lighting, surround, and observer panel will make a difference in the result also.

This indicates that there are several useful principles that can be employed when developing a scaling plan for a perceptual color space. This fact and the fact that the system creator may want to incorporate additional principles into the sampling plan indicate the potential for several different kinds of systems. If they have been prepared with equal care, the choice becomes one of a field of application and preference. In general, it is possible to express the facts of one system in terms of another, and several such comparisons have been published for a number of well-known systems. The situation is somewhat comparable to choosing a system of measurement: arguments can be made in favor of the metric or the foot/pound/gallon system, and results in one can be converted to results in the other. The historical high interest in scales of equal perceptual magnitude is due to potential use in objective color control of manufactured goods. Before discussing two attempts at a uniform color space, a system based on color-magnitude judgment will be described.

SWEDISH NATURAL COLOR SYSTEM

NCS is a modern example of scaling of perceptual color space conceptually involving color magnitudes. It is based on Hering's ideas of composition of color experiences from fundamental color (with unique hues), whiteness, and blackness experiences. The developers of NCS claim that color-normal observers can judge with good accuracy the magnitudes of one or two unique hues, whiteness, and blackness in a complex color experience. Their magnitude in a color chip can be judged based on innate concepts of the components. However, in Chapter 4 significant variation in color chips identified as having unique hue, particularly green and red, was mentioned. Similar variation has been obtained in judgments of black and white magnitudes. Independent tests of magnitude judgments of NCS chips have not resulted in the level of reproducibility claimed by the system developers. We can consider the system to be representative of an undefined "standard" observer producing judgments implicit in the system (11).

Its coordinate system represents the four unique hues, as well as whiteness and blackness. The components always add up to 100, that is, they can be taken as percentages of the total perceived color. Colors of a given hue are located on an equilateral triangle with white, black, and the full color at the corners (Fig. 5.7a). Tint and shade scales run from the full color (C) to white (W), respectively, black (S). The gray scale connects the latter two. Colors located in the interior of the triangle contain C (where C consists of one or two neighboring unique hues) as well as W and S. If, for example, the values of full color and blackness are known, the value for whiteness can be calculated. Degree of chromaticness is indicated by the vertical lines parallel to the gray scale. In the chromatic plane there are nine grades between two neighboring unique hues (Fig. 5.7b), resulting in 40 steps around the hue circle. Each triangle specifies 66 colors, resulting in over 2000 aim-color specifications for the atlas of the system. Of these, 1741 have been realized in the form of painted paper samples (limited to this number because of lack of suitable colorants). These samples represent the aim colors only under standard conditions of illumination and surround. The system fits into a Euclidean space in the form of a double cone and makes no claims for perceptual equality of steps.

NCS includes a color identification system. A color is identified, for example, as 4030-R70B. This indicates a color of blackness $s = 40$, chromaticness $c = 30$, and a hue consisting of 30 parts unique red and 70 parts unique blue.

The atlas represents a specific implementation of Hering's ideas. While tools of color technology have been used for the production and control of the atlas samples, the concepts are taken to be universally applicable and rely on the observer only. For this reason the system is popular with nontechnological users such as designers or architects and is used in situations calling for color definition where instrumental measurements are difficult. It is important to realize that in such situations a given material can result in different color perceptions, depending on reflectance function, illumination, and surround colors.

MUNSELL COLOR SYSTEM

Equality of difference in color-order systems has been an elusive goal of color-order system developers since the eighteenth century. It was the American artist and educator Albert Munsell (1858–1918) who initiated the first attempt at such a system at the beginning of the twentieth century. Munsell also departed from the idea of placing the most saturated colors, regardless of hue, on a common plane. He introduced lightness (using the painter's term value) as one of the three primary attributes. He

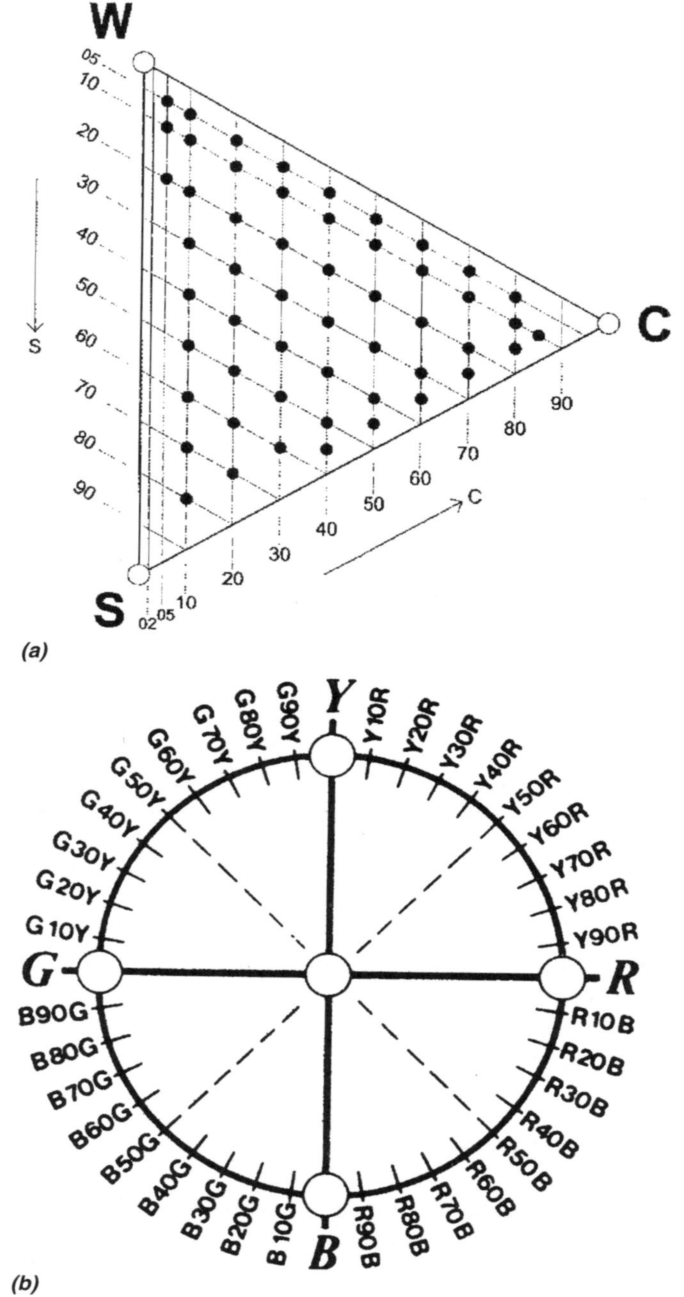

FIGURE 5.7 *(a) Organization of the NCS system (11), constant hue triangle with white W, black S, and full color C. Dots represent color samples identified according to blackness s and chromaticness c. (b) Hue circle with average unique hues on the semiaxes.*

discovered quickly that different pigments investigated by him resulted in color chips with different maximum chroma. He came to understand from these facts that a uniform color space could not fit into a simple geometric space. Munsell invented the term chroma for chromatic intensity of object colors and made its scale open-ended. A wish for perceptually equal steps in all three attributes clashed with the desire to have round numbers of hue and value grades. The decision to have 10 value and 100 hue steps had the consequence that the step sizes, even though uniform within an attribute, were of a different magnitude in the two attributes (12).

The Munsell system remains today the best-known global color-order system. Its internal structure is cylindrical, with the hue attribute represented by radial lines originating at the central axis. Lightness is represented by height in the cylinder and chroma by concentric circles around the central axis (Fig. 5.8a–5.8c). Munsell liked the decimal system and used five primary hues in his conceptual 100-hue circle (of which he originally sampled only 20, later 40): yellow, red, purple, blue, green. If we consider the modern Munsell hue scale to be uniform (which it approximately is), we find different numbers of steps between average unique hues:

Unique hue sector	Munsell hue steps
Red to yellow	20
Yellow to green	23
Green to blue	26
Blue to red	31

This is an indication that the hue circles of the NCS and the Munsell systems are distinctly different. In the Munsell system, the number of unit hue differences between colors of neighboring hues depends on chroma, as mentioned earlier. It is larger at high chroma than at low chroma.

The chroma scale is open-ended, starting at zero chroma in the center and increasing radially. Practical limits are set by the maximum chroma of available pigments. The theoretical limits are reached at the surface of the object color solid, representing ideal optimal colorants (see Fig. 6.11). The lightness scale has 100 grades, of which 10 are available as samples in the atlas of the system. Experimental evidence of lightness crispening was smoothed out in the final lightness scale. It is identical for all colors regardless of hue and chroma, and thereby does not represent perceived lightness because the Helmholtz-Kohlrausch effect has not been considered (see Chapter 4). Editions of the atlas with increasing numbers of samples were published in 1906, 1915, and 1929. The sample colors for these atlases were arrived at using visual scaling of Maxwell disk colors and of color chips. In 1943 a committee of the Optical Society of America made recommendations for revised aim colors of the Munsell system defined in the Commission International de l'Éclairage (International Commission on Illumination, CIE) system of colorimetry, the so-called Munsell Renotations (13). The Renotations continue to be the specifications for the commercial system. The committee defined 2746 chromatic and 9 achromatic colors reaching to the theoretical limits. In the commercial system, approximately 65% of these have been realized as color chips in matte and glossy editions. Care has been taken in modern pigment

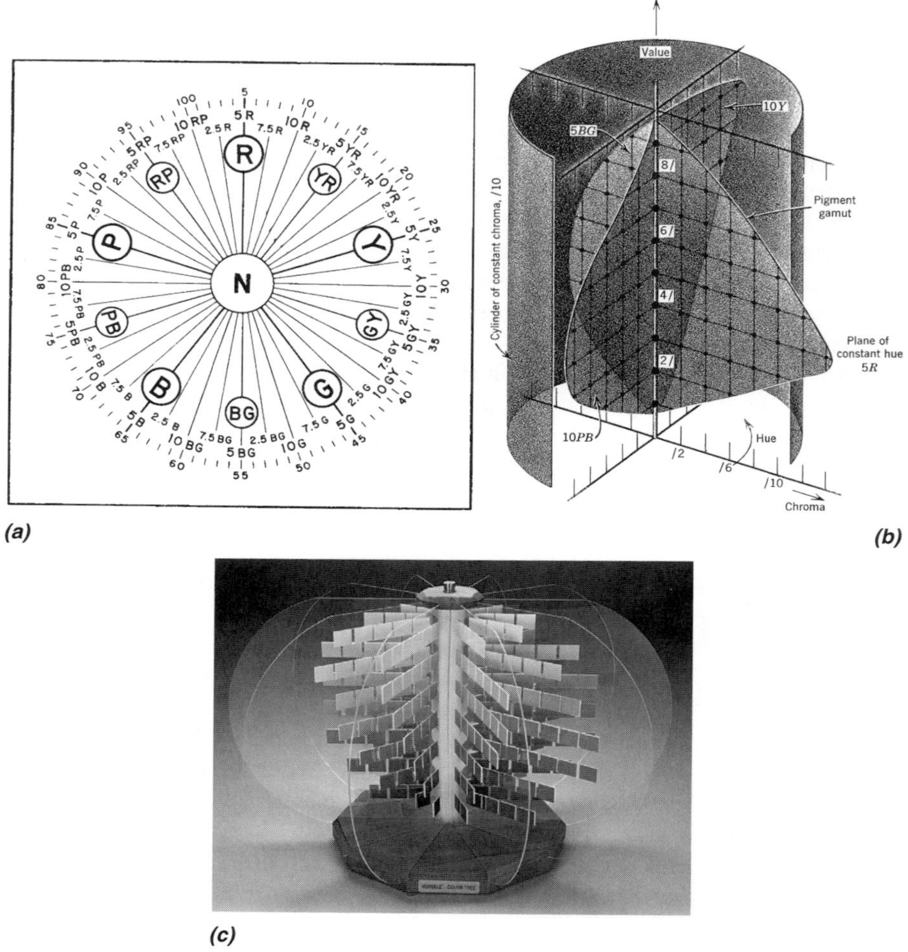

FIGURE 5.8 *(a) Organization of the Munsell color chart. (b) Conceptual illustration of the organization of the Munsell system (12). (c) Model of the Munsell Color Tree. (Image courtesy Gretag-Macbeth Corp.)* Figure also appears in color figure section.

formulations of the system so that colors appear relatively constant when the chips are viewed in different natural and artificial versions of daylight. The surround conditions for which the system is considered accurate have not been defined. Interestingly, the 1906 edition contained plates of the value 3 plane both against a white and a black background, illustrating the dramatic change in appearance of the chips. This comparison was eliminated in later editions.

The Munsell system includes an identification scheme for colors. A complete designation is in the format hue/value/chroma, for example, 5PB 6/12 representing a color of hue 5PB (purplish blue), value 6, and chroma 12. Intermediate values can be interpolated, and computer software is available for this purpose. The popularity of the system rests in part on the good level of comprehensibility, even for untrained persons, of the three attributes of the system. However, some people have difficulty in distinguishing between lightness and chroma steps of unsaturated colors.

The limitations of the Munsell system as a uniform global color order system have resulted in an effort of nearly three decades by the Committee on Uniform Color Scales of the Optical Society of America to develop an improved uniform, isotropic system.

OPTICAL SOCIETY OF AMERICA UNIFORM COLOR SCALES

If a color space divided according to a polar coordinate system, such as the Munsell system, cannot represent perceptual uniformity in all directions, is there a three-dimensional geometry that can? This is the fundamental question the Committee on Uniform Color Scales attempted to answer. Hue/chroma orientation of the chromatic diagram was not the answer. A geometrical solid with equal distances in all directions from the center point (that is, representing all possible colors different by unit perceptual distance from the reference point) is a sphere. But a sphere cannot be packed solidly into a space without gaps or overlaps. The regular geometrical solid, offering at least twelve directions of constant distance while at the same time packing without gaps is the cubooctahedron (Fig. 5.9). That this is the case is illustrated in Figure 5.10, which shows double expansion of a cubooctahedron in all three dimensions.

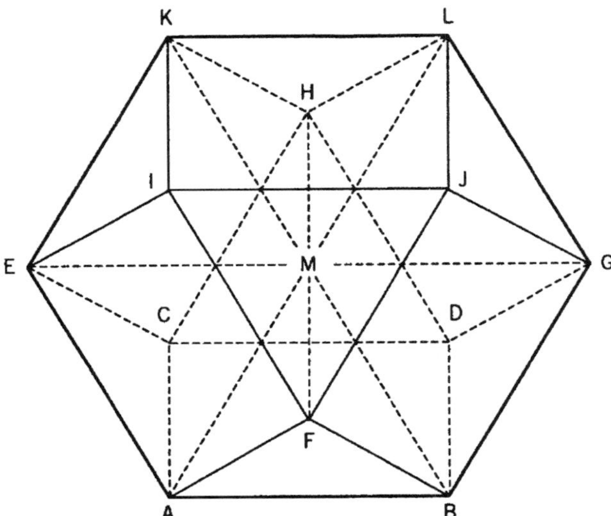

FIGURE 5.9 *Cubooctahedron as the organizing principle of OSA-UCS.* M *represents the central reference color, twelve equally distant colors are identified by the other letters.*

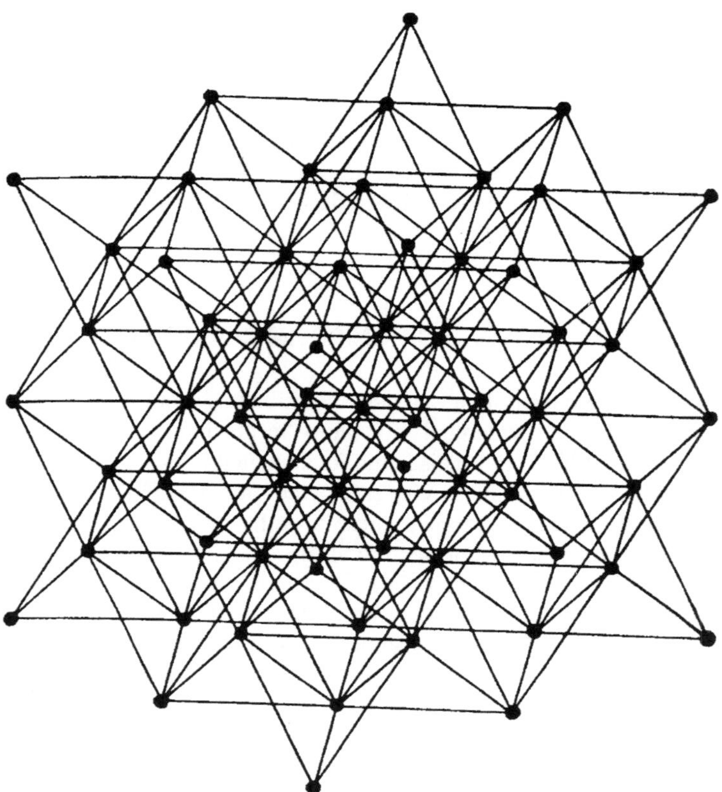

FIGURE 5.10 *Cubooctahedron doubly expanded in all directions (14).*

The resulting crystalline structure makes possible a space that is geometrically uniform in 12 directions, and if it can be filled appropriately with color samples, comes closest to an isotropic color space. This is the approach taken by the committee in its experimental work.

In a given direction, the grid of points separated equally is hexagonal, and thus triangular (plane A, B, G, L, K, and E in Fig. 5.9). The committee prepared a series of 43 samples arranged in a triangular grid pattern at approximately equal lightness. The perceptual distances of 107 resulting pairs of samples (having mostly mixed hue and chroma differences) were compared against a gray surround by 76 observers and the average perceptual distances between samples calculated from the results. The results, unsurprisingly, indicated that a uniform chromatic diagram in a flat plane could not be obtained. Calculations showed that the unit perceptual contours were elongated at an approximate ratio of 2:1 and that hue superimportance, therefore, applied. The committee decided to "Prepare a set of color chips representing the closest a Euclidean system can come to a uniform color solid . . . " (15,16). Additional tests were performed to establish the relationship between unit chromatic and lightness differences and to determine the magnitude of the HKE. A mathematical formula was fitted to the experimental data and the formula was used to calculate aim points in the cubooctahedral pattern. Paint formulations were then established for the calculated points. A result of the cubooctahedral arrangement is that within the solid there are seven flat cleavage planes, within which the perceptual differences are equal (discounting the hue superimportance effect). In terms of the cubooctahedron of Figure 5.9, the constant lightness plane is formed by colors E, F, G, and H. The other six planes connect the following colors:

Plane 2	B, F, I, K, H, D
Plane 3	A, F, J, L, H, C
Plane 4	I, J, G, D, C, E
Plane 5	K, L, G, B, A, E
Plane 6	I, A, D, L
Plane 7	J, B, C, K

These planes slice through the complete color solid and open views of colors that are approximately equally distant from neighboring colors (see Fig. 5.11, for some examples). The vistas opened by these sections are a consequence of the geometrical structure. Artists and designers have studied such cleavage planes for esthetically pleasing and harmonious color selections. The geometrical model of Fig. 5.11 illustrates the 424 colors of the regular set. On studying the set, committee members saw a need for more nearly neutral colors, and 134 color samples were calculated for this subset and prepared. The atlas issued by the Optical Society in 1977 therefore contains a total of 558 glossy color chips.

The committee also devised a color identification method based on the system. Lightness, designated L, has a value of zero at the lightness of the surround gray and ranges from -7 for the darkest colors to $+5$ for the lightest. There are two chromatic designators, j (for the French *jaune*, yellow), indicating roughly yellowness, respectively, blueness, ranging from -6 (saturated blue) to $+12$ (saturated yellow), and g for approximate redness, respectively, greenness, ranging from -10 for red to $+6$ for green. The "pastel" subset is in half-steps from $L = -1.5$ to $+1.5$.

The adjustment of the experimental data to fit a Euclidean space is not widely known, but explains the apparent discrepancy between Munsell and OSA-UCS data. Without the intuitive attributes of hue and chroma, it takes practice to find one's way around the system. As a result the OSA-UCS system has not, as initially expected, taken the place of the Munsell system.

ARE LARGE-SCALE GLOBAL COLOR SCALING DATA SETS IN AGREEMENT?

Comparison of different color scaling data is best done by comparing the stimuli that represent the samples. In Figure 5.12 constant chroma circles, as determined in large-scale experiments (over 10,000 observations) and related constant difference

FIGURE 5.11 *Model of samples of OSA-UCS, illustrating some of the cleavage planes of the system. (Image courtesy D. L. MacAdam).* Figure also appears in color figure section.

hue scaling, are plotted in a common diagram. The three sets of data compared are Munsell Renotations, a preliminary experiment by Nickerson and co-workers in connection with development of OSA-UCS, and data from the so-called Munsell Re-renotations, where the key findings behind the major OSA-UCS experiment have been translated into a revised set of Munsell aim colors. In the three sets of data, the size of the unit chroma difference varies somewhat. The key point is that the shape of the unit chroma contour and the spacing of the unit hue differences along that contour vary significantly in the three experiments. Different methodologies, possibly different surrounds, and different observer panels have been used in the

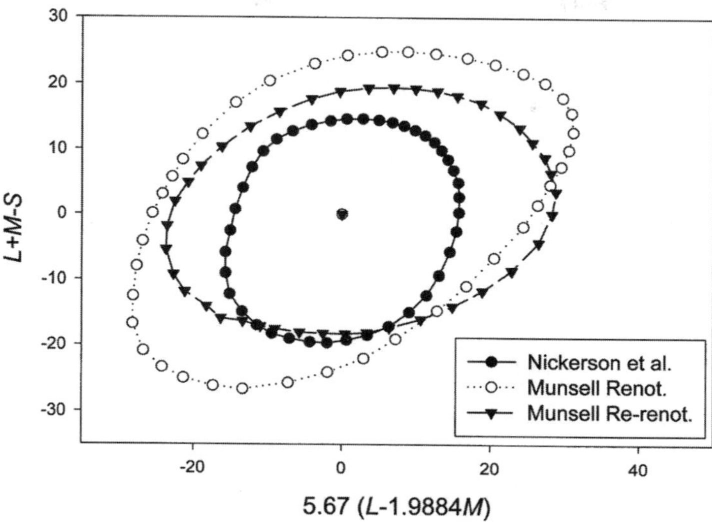

FIGURE 5.12 Constant chroma contours and 40 peceptually equally spaced hue differences of three large-scale experiments in a common, cone-based, balanced chromatic diagram: Munsell Renotations, Munsell Re-renotations, Nickerson et al.

three experiments. No claim can be made that one set of results is more accurate than the others, and the reasons for these discrepancies are unknown. The results indicate that it is difficult to establish perceptual color scaling data with a high degree of general reliability. Comparable variation is also found in small color difference data and the degree to which such data can be replicated is unknown at this time.

Another indication of the degree of agreement (or lack of it) between major data sets is found in Figure 5.13, which illustrates a plot of the OSA-UCS $L = 0$ chromatic plane in the Munsell perceptual diagram. For good agreement the OSA-UCS grid would be uniform and evenly spaced (17).

Some of the key findings in this chapter indicate that reliable color scaling is difficult for reasons that are not yet understood. The considerable variability of color-normal observers in color stimuli picked as expressing their unique hues is one possible reason. Overlooked differences in methodologies with a significant effect on the results may be another.

Another important conclusion is that a perceptually uniform color space cannot be accurately represented by a Euclidean geometric model because of hue superimportance. It is possible to represent color spaces in many forms of simple geometrical space, but the resulting constant interval spacing has little or no direct connection to psychologically meaningful scaling. An intermediate step is represented by the NCS system. Here arbitrary decisions have been made to place full colors in the same plane and assign the same saturation value to them. As a result they fill a double-cone space. In terms of perceived differences its chromaticness scales vary by hue, and the vertical dimension does not represent lightness, but rather is without definition.

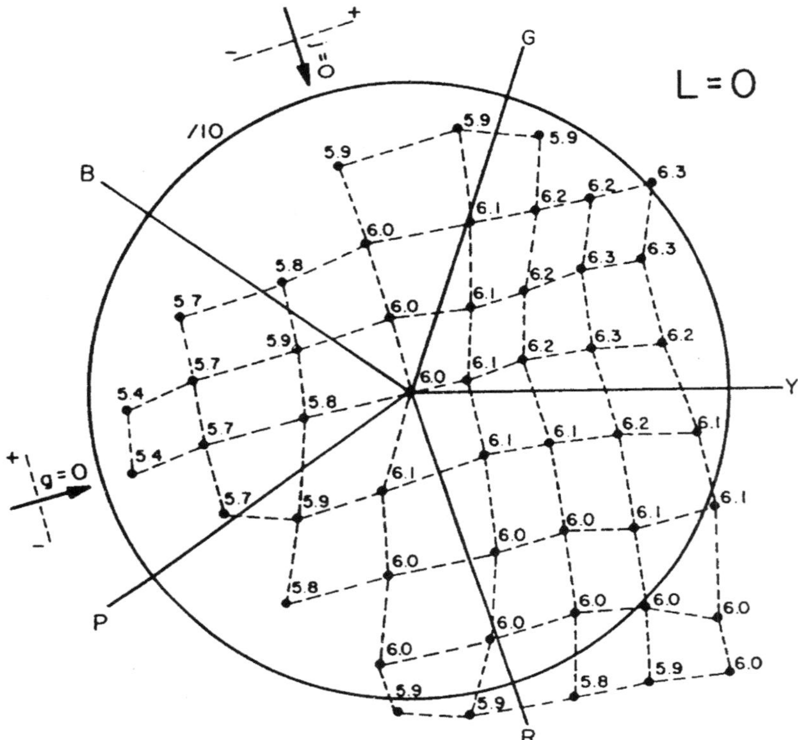

FIGURE 5.13 *Plot of OSA-UCS samples at* L = 0 *in the Munsell chromatic diagram illustrating the degree of agreement between the two systems (15). Directions of the principal hue axes* g *and* j *of OSA-UCS are indicated by arrows. The numbers represent Munsell values of the OSA-UCS samples.*

OTHER COLOR-ORDER SYSTEMS

Dozens of other color-order systems have been developed over the last 200 years, many short-lived. Well known among these is Ridgway's color atlas of 1912, or Ostwald's color atlas with 2500 samples (18). It was issued in the United States in three editions as the *Color Harmony Manual*. In Germany a standard color-order system, DIN 6164, was developed in the middle of the twentieth century (19). It is ordered according to the attributes hue, saturation (different from chroma), and relative darkness, and continues the European tradition of placing full colors on the same plane. As the Munsell system, it is only uniform in some respects.

The recommendation by the CIE of the CIELAB color space and color difference formula for object colors in 1976 has resulted in the development of several systems based on that formula. First was the *Eurocolor* system, no longer available. Other European versions are the *RAL Design System* and the *Acoat Color Codification* (*ACC*) system. In the United States the *Colorcurve* system, designed by Stanziola, was introduced in the early 1990s (19). It has four elements: systematic (but irregular) aim points in CIELAB space, physical samples representing them, reflectance data for each aim point suitable for computer-assisted formulation as well as colorimetric data, and computer software to use the system efficiently. The atlas has two volumes: the master atlas with 1229 samples at 18 lightness levels in regular grids on the a^*, b^* diagram. The second Gray and Pastel Atlas contains 956 light and grayish color samples in smaller increments. There is a color identification system expressed with the following format example: L65 R4Y3, where the first figure indicates the lightness of the sample as the L^* value, and the following letter/number combination indicates the position on the chromatic grid represented by the four CIELAB chromatic semiaxes. Because it does not consider hue superimportance nor the HKE or crispening, CIELAB does not represent an isotropic color space, and therefore neither do these systems.

There are several kinds of colorant and color-mixing-order systems. Among these are printing ink systems in halftone printing, usually in cube form, such as the one patented in the United States in 1969 by Wedlake, the *Küppers DuMont Color Atlas* with 5500 printed colors of 1975, and the modern cyan, magenta, yellow, black (CMYK) system used in color printing (21). These are systematic colorant mixture systems with only ordinal connection to perceptual scaling.

Additive (light) color-order systems are used in connection with computer displays. The two most commonly used versions are HSB (hue, saturation, brightness; cylindrical in form) and RGB (based on the three additive primary colors red, green, and blue produced by the phosphors of the display unit; cubic). The widely used software packages Adobe Illustrator® and Photoshop® use a cubic display in the RGB mode, the Color Picker, to demonstrate available colors. Colors can be specified or selected according to HSB, RGB, CIELAB, or CMYK scales, depending on the need. These choices result in different display formats. These systems are based on increments of color stimulus and have no (or only limited in the case of CIELAB) connection to perceptual scales.

SPECTRAL SPACES

So-called spectral spaces are another kind of space related to color stimuli. Color stimuli, in the normal case, are spectral power distributions absorbed by the retina in the eyes. Visible light ranges approximately from 400 nm to 700 nm, as discussed in Chapter 1. At 10-nm intervals this equals 31 spectral values, or 31 dimensions. The three cone types in the retina reduce these dimensions by filtering to three. The three dimensions can form a color solid, either directly or with modification. The spectral sensitivity of the cone types probably represent a compromise between what is biologically possible and what was evolutionarily most beneficial to our early ancestors at a time when our vision system developed in its modern form. We cannot assume that this compromise represents the perfect mathematical answer to recovering the information contained in the spectral power distributions (see also Chapter 2). There are several mathematical methods for such dimensionality reductions that optimize the reduction process in different ways (22). One of these is principal component analysis (PCA) used on color reflectance data for the first time by Cohen in 1964 (23). Figure 5.14 illustrates the results of the first three principal

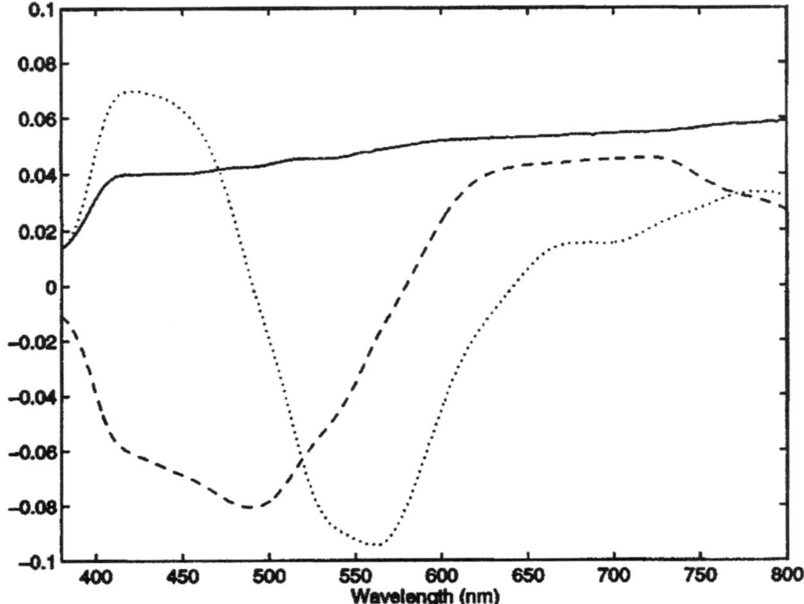

FIGURE 5.14 The first three principal component functions derived from analysis of the spectral reflectances of 1269 Munsell color chips. The second function (dashed line) has some resemblance to the yellowness–blueness opponent color function, the third (dotted line) to the redness–greenness function (24).

components of the complete set of Munsell chip reflectance functions. There are interesting rough similarities to the luminosity and the two opponent color functions, indicating that the purpose of the cone functions may be the approximate recovery of the spectral power function of the stimulus. PCA with three functions recovers reflectance functions of the Munsell set with notably higher accuracy than the cone functions (ca. 97% vs. 90% for color-matching functions). For recovery in the 99+% region, typically five PCA functions are required.

A problem with PCA and similar functions is that different data sets result in somewhat different functions (depending to some extent on the colorants used in their manufacture). Thus different subsets of Munsell reflectances or the NCS reflectances result in different PCA functions than the full Munsell set. When PCA results of Munsell data are plotted in a space of the first three functions, they are found to be in ordinal hue and chroma order, but their distances are not related to perceptual distances. Like cone functions PCA and other dimension reduction functions result in metamers. However, metamers resulting from the different methods vary.

COLOR NAMING

Once colors are arranged in an orderly fashion they can be named. How color names developed historically is a complex issue. There are two theories, one stating that people of all societies become aware of different colors or color categories and then named them in essentially the same sequence: white and black, red, green, yellow, blue, brown, purple, pink, orange, gray (25). The other theory is based on the belief that all color names are group cultural achievements and there is little common thread. As in the wider nature vs. nurture controversy (of which this is a segment), the truth likely is somewhere in the middle: it is likely that there is a genetic element to color naming, but in every specific situation a smaller or larger cultural component can be expected. The first four chromatic colors in the list are those of the Hering opponent color theory, the most distinct hues possible; brown is ubiquitous in nature. The simple one-syllable names for the first seven of these colors have similar roots in some languages, indicating that they are very old. For example, the Indo-Germanic word for red is believed to be *reudh*, the basis of the Greek *erythros*, Latin *ruber*, German *rot*, French *rouge*, Spanish *rojo*, and so on.

Many other color words are related to materials, such as sea green, orange, ultramarine, olive, malachite green, chartreuse, and the like. Other words are compounds such as bluish red, or have qualifiers attached such as dark, light, vivid, and so forth. Some names reflect poetic invention, like Cuban Sand, Ashes of Rose, Blue Fox, and so on. As discussed earlier, several color-order systems have color identification schemes attached to them. For nonexperts they are not very descriptive. A simple scheme that allows naming of colors with some easily recognizable specificity is desirable. Such a system was proposed for the English language in 1955 by the then U.S. National Bureau of Standards. It has six levels, from the least precise level 1 to the highest precision at level 6. The first level consists of thirteen color names, the eleven just mentioned, plus yellowish green and olive. At level 2 sixteen intermediate hue names are added, such as reddish brown or bluish green. On level 3 there are 267 subregions identified, their central colors later demonstrated as color chips in the *ISCC-NBS Method of Designating Colors* (25). A typical example is "light yellowish brown." The modifiers include light, strong, brilliant, vivid, dark, deep, very, and so on. However, an atlas is required to assign these names uniformly. For many practical purposes, such as in archeology, botany, or to identify bird and insect colors, this level of detail is satisfactory and the terms are easy to understand. Level 4 has approximately 5000 possible designations. It is based on existing sample collections, such as the Munsell system, and uses their designations. Level 5 uses interpolated or extrapolated Munsell notation to arrive at some 100,000 designations. Finally, level 6 uses colorimetric stimulus specification, such as the x, y, Y values of the CIE colorimetric system (see Chapter 6). The number of colors distinguishable at this level is without limit. At higher levels, designations become increasingly abstract and more difficult to comprehend.

It is evident that color order, at least in form of unit perceptual differences, is a complex endeavor. In the next chapter a quantitative description of color stimuli is discussed as a basis for attempts at mathematical models of isotropic color order.

6

Defining the Color Stimulus

A belief that there might be a close agreement between color stimulus and our color experiences is, on the surface, reasonable. It follows general experiences we have that indicate reactions are immediately due to certain actions, and in some cases they are. However, in the case of color perception, the action is contact of quanta of light with cone cells, and the reaction is a subconscious and conscious interpretation of the scene in front of our eyes with a very complex but sufficient apparatus supporting our survival. There are countless examples known, indicating that there is no simple one-to-one relationship between stimulus and experience.

In color technology it has been established that by simplifying the total stimulus as much as possible a useful degree of correlation between stimulus and response can be obtained, that is, surround, illumination, and test procedure are controlled in some fashion, and data for an average observer are determined. This is the basis of the (limited) success achieved in this field.

On the simplest level the physical stimulus from a light is defined by its spectral power distribution (SPD) and its intensity (1). Since, within a range around average daylight light intensity, perceptual results are not much affected by intensity, SPD is taken as sufficient for many technical purposes. The stimulus, under comparable conditions, from an object is defined as the light from a standard light source as modified by the reflectance properties of the object and expressed in terms of an SPD (spectral return). There are immediate technical problems in that the SPD of daylight can vary considerably as a function of time of day and weather. Similarly, there are many kinds of lamps, and the SPD of the light they give off depends on type of lamp, manufacturing components, and age of lamp. In practice, close standardization

Color: *An Introduction to Practice and Principles, Second Edition*, by Rolf G. Kuehni
ISBN 0471-66006-X Copyright © 2005 John Wiley & Sons, Inc.

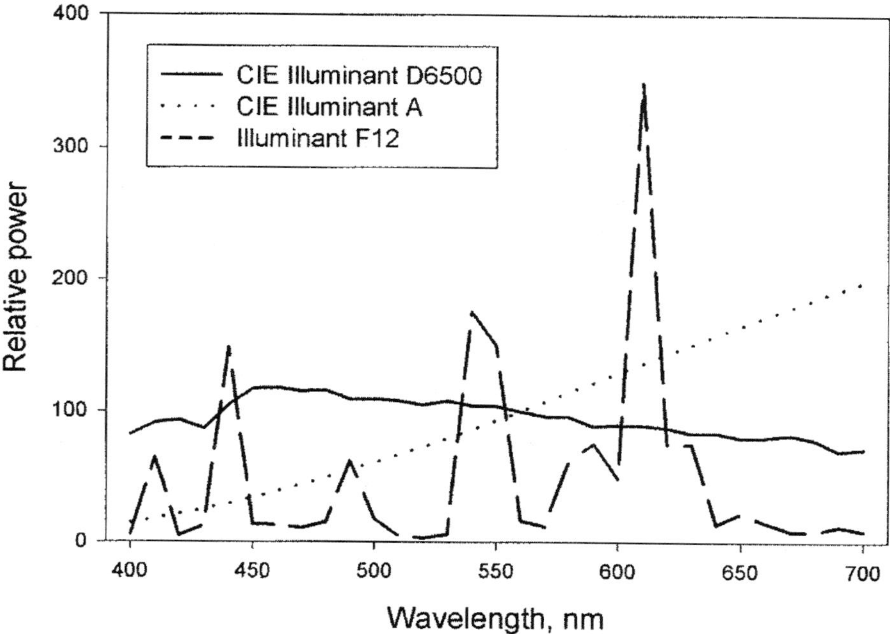

FIGURE 6.1 *Spectral power distribution functions of CIE daylight D6500, incandescent light A, and F12 triband fluorescent light.*

is difficult. Further, there are no artificial daylight lamps with SPDs in close agreement with actual daylight SPDs. This problem is somewhat reduced in importance because of the adaptation process, but not in an easily predictable way. In this matter the approach taken by the CIE was to specify for the purpose of calculations several standard illuminants with sets of numbers. Different standard daylight SPDs are classified by their correlated color temperature on the Kelvin scale (see Chapter 1), that is, by the temperature of a blackbody giving off light of the same chromatic (not spectral) properties. In tests lights of about 4000 K have been found, on average, to appear neither yellowish nor bluish. Before adaptation, light sources with higher Kelvin numbers appear increasingly bluish. The most commonly used standard illuminant is D6500 (or D65) representing northern sky light. Tungsten lamp light (common light bulb) has been standardized as illuminant A, with a correlated color temperature of 2500 K. Several types of lights from fluorescent lamps have also been standardized (F illuminants). Figure 6.1 illustrates three typical light source SPDs (2). As can be seen, their spectral forms are distinctly different. The spectral power of lights is measured with a spectroradiometer. These instruments are comparatively expensive and, in practical technological circumstances, SPDs of actual light sources used in visual evaluations are rarely measured, even in color difference scaling experiments. Light sources therefore represent a variable in stimulus specification.

Spectral reflectance functions of objects are measured with a spectrophotometer. Figure 6.2 is a schematic sketch of such an instrument. The CIE has specified two

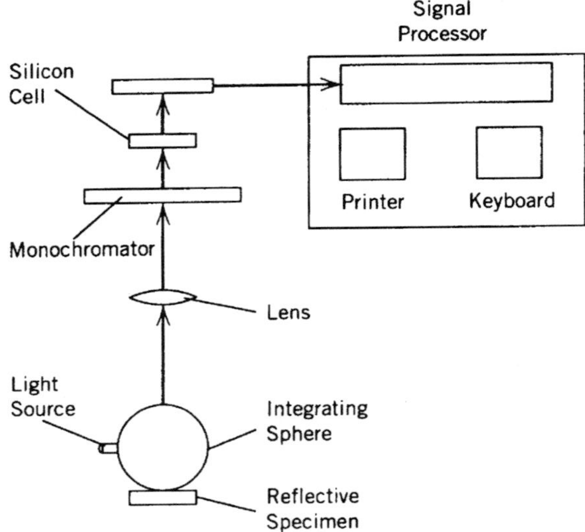

FIGURE 6.2 *Schematic representation of the components of a spectrophotometer.*

basic standard measuring geometries for these instruments: 45°/0° and its reverse, and diffuse/0° geometry and its reverse. In the former type, light strikes the object at a 45° angle and its reflection is measured at 0°, that is, perpendicular to the object. The latter case takes account of the fact that surfaces of many objects are more or less uneven, for example, those of textile fabrics, and incident light is scattered in many directions. To obtain an average reflection value the sample is attached to an integrating sphere, a hollow sphere with an interior white coating. Light scattered from the material averages out in the sphere and is sampled at a particular angle. Reflectance functions are relative values of the ratio of incident to reflected light, and thus not dependent on the SPD of the light used to make the measurement. There are two exceptions: (1) fluorescent colorants (see Chapter 8), and (2) if the amount of light from a given source is very limited at certain wavelengths and thus affects the accuracy of measurement. Figure 6.3 illustrates the spectral reflectance functions of three objects.

This brief discussion indicates that specifying the reflectance function of a material is no trivial matter and that, depending on the conditions of the measurement, significantly varying results can be obtained. International and national standardizing committees are involved in continuing efforts to improve reliability of such measurements. A physically exact definition of light stimuli arriving at the eye is more complex than what has been described here. However, for technological purposes separate determination of reflectance function and spectral power distribution is considered adequate.

The relative "standard" stimulus can be calculated from reflectance measurements and light source SPDs by multiplying, at each wavelength, one with the other. Figure 6.4 illustrates the relative spectral forms of the stimuli arising from illuminating the same object with three different standard illuminants, all producing white-appearing light.

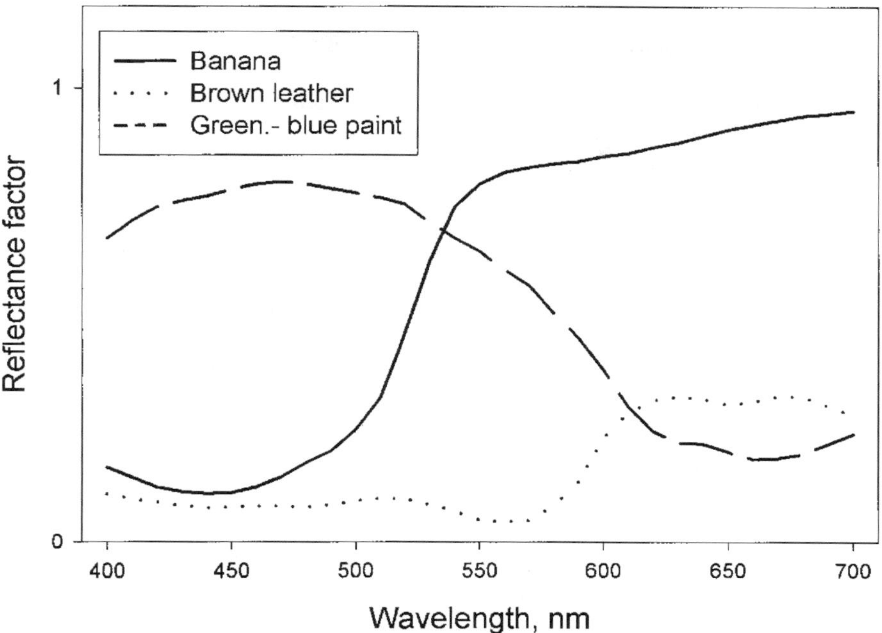

FIGURE 6.3 *Spectral reflectance functions of three objects.*

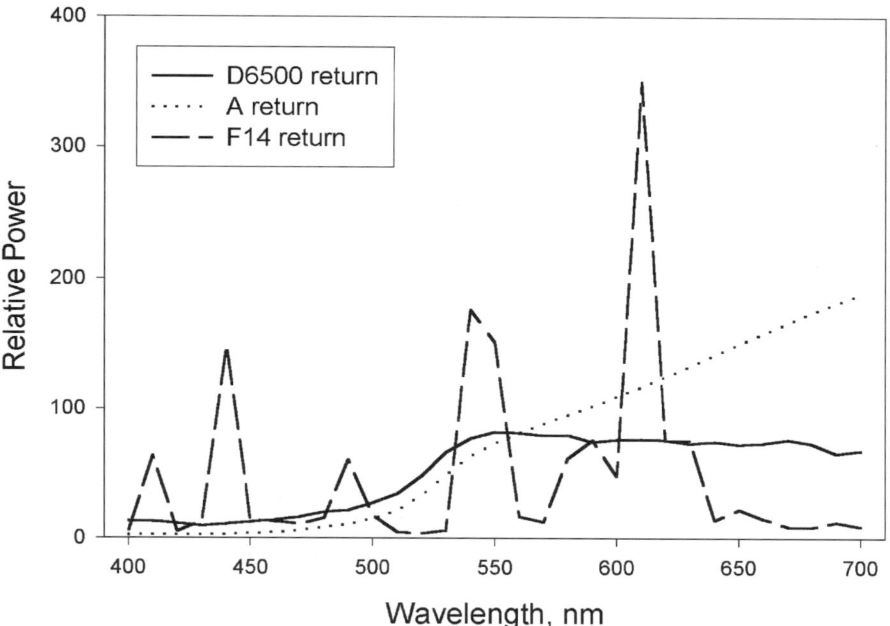

FIGURE 6.4 *Relative stimuli from the banana of Figure 6.3, as returned from lights with the spectral power distributions shown in Figure 6.1.*

MATCHING STIMULI

As mentioned earlier, in the visual process, light stimuli are filtered with cone sensitivity functions. As a result, the spectral complexity of stimuli is reduced to one value each for the three cone types that represents the result of the filtering process. Such numbers are deemed representative of the average output of a group of cones on which the image of the stimulus falls. From mid-nineteenth century on, color fundamentals (cone sensitivity functions) were determined with the help of color-matching experiments. The appearance of a spectral light was matched with an appropriate mixture of two or three other spectral lights. The test subject views a circular field, separated into two parts. In one part the reference wavelength is displayed, in the other the intensities of the three standard wavelength lights can be adjusted until equality of appearance (a match) is obtained. By doing this in systematic fashion with a suitable piece of optical equipment (a visual colorimeter), functions can be calculated from the results that directly or indirectly represent the cone sensitivity functions of observers (3). Already in the first experiments of this type, noticeable differences between individual observers were obtained (in addition, observers were found that had only two, one, or no cone function).

Such matching is based on the principle of metamerism (see Chapter 4). Under matching conditions the total impact of three spectral lights on the cones is the same as that of the single light, that is, the cone filter values are identical. In his classic experiments of 1931, Guild used lights of wavelengths 460, 540, and 630 nm to match the appearance of all spectral lights (at 10-nm intervals) (4). The results for seven observers are illustrated in Figure 6.5. Their values for the three lights are represented by the curves. At the three wavelengths of the standard lights, the values of two lights are at zero, while the third has a value of one. Here the reference light was matched with itself. A surprising fact is that at certain wavelengths the amounts of one of the standard lights is negative. This is proof that with light of three wavelengths it is for color-normal observers, not possible to match all spectral lights. While the hue can be matched in all cases, saturation of the test mix is often reduced compared to that of the reference. To obtain equality of appearance, a certain amount of one of the standard lights has to be added to the reference light. In the metameric "equation" this light is considered negative. The three spectral functions have conventionally been named color matching functions $\bar{r}$, $\bar{g}$, $\bar{b}$. They do not have the form of the cone response functions, but one that is presumably mathematically linearly related, as will be discussed later in the chapter.

It is important to note at this point that by doing this experiment nothing has been established in regard to appearance. The functions simply indicate for a given observer combinations of standard lights that match all the spectral lights in appearance without indicating the qualitative character of the appearance of the lights. To compare the results of different experiments, they had to be appropriately normalized.

In 1931 the CIE met to consider a proposal for standard observer data representing average results of color-matching experiments using a 2° field of view. Two issues needed to be resolved: (1) all three functions had negative values that made computation (given the available equipment at the time) cumbersome, and (2) none of the

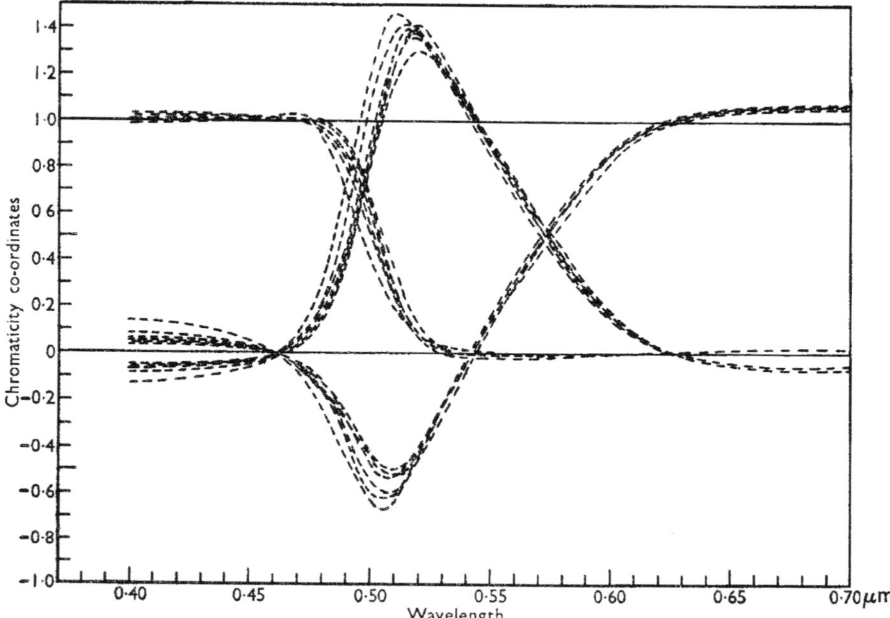

FIGURE 6.5 *Results of metameric matching experiments of the appearance of spectral lights using mixtures of lights of wavelengths 460, 540, and 630 nm. The results of seven observers are plotted individually, showing the amounts of the three standard lights required at each wavelength (see Note 3 for reference).*

three color-matching functions represented brightness perception, and the question was how to include brightness perception in the system. A proposal was made to linearly transform all three functions, so that one of them was identical to the CIE spectral luminance function adopted in 1924. Avoiding negative colors can result in functions that represent cone sensitivity. But having one of the functions agree with the standard luminance function would make at least two of the three functions different from cone sensitivity functions. Here it is useful to insert a brief discussion on linear transformation.

Matches of lights are sometimes expressed in terms of mathematical equations even though it can be argued that adding three lights is very different from adding the value of three coins, for example. However, in the mid-nineteenth century, Grassmann predicted, on the basis of some assumptions about the mathematical treatment of lights, certain results that were soon shown by experiment to be valid (5). The resulting laws imply that if the results of matching experiments with three given standard lights are known, the corresponding results for three other standard lights can be calculated by a process of linear transformation. To go into the details of this procedure exceeds the bounds of this text, so the interested reader is referred to any text on solving systems of linear equations. Experimental results have shown that the cone system, when tested in the reduced circumstances of the light-matching experiment, follows

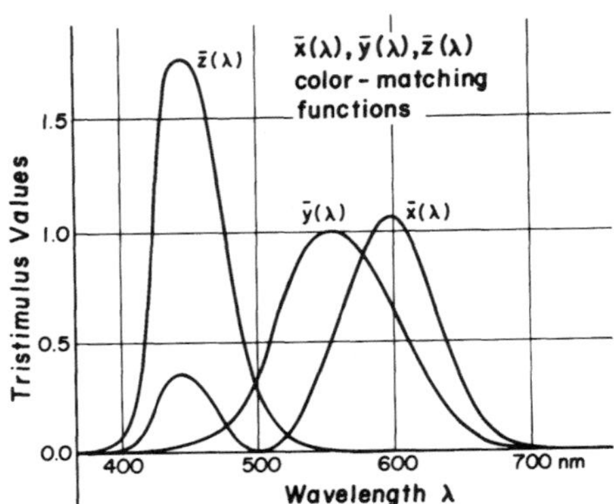

FIGURE 6.6 *Spectral color-matching functions of the CIE 1931 2° standard observer (8).*

Grassmann's laws quite well (6). Given this fact, linear transformation has been used to convert experimental r, g, b, color-matching functions to cone sensitivity functions and to a special set of functions that the CIE promulgated as CIE 1931 2° standard observer functions (7), used in its system of colorimetry. The functions are identified as $\bar{x}$, $\bar{y}$, $\bar{z}$, and the $\bar{y}$ function is identical to the CIE luminance function of 1924. These functions are illustrated in Figure 6.6. It is important to note that there are no real lights that correspond to these three functions. They are the results of mathematical conjecture. These functions allow determination not only of what single wavelengths and mixtures, but also which broadband stimuli, such as those of object colors, are metameric for the standard observer.

THE CIE COLORIMETRIC SYSTEM

Specification of standard methods for defining illuminants (SPD), the reflectance properties of objects, and the spectral sensitivity of average human cones makes it possible to place any color stimulus into a three-dimensional space. In the CIE colorimetric system, the coordinates of the space are formed by the so-called tristimulus values. For object colors these are obtained by multiplying at each wavelength the relative SPD of the illuminant with the reflectance values of the object and, in turn, with the spectral values of the three color-matching functions. The result represents the stimulus as absorbed by the three cone types (in linearly transformed form) and the area under the curves is integrated or, more practically, for a given tristimulus value the individual values at each wavelength are added up. This process is illustrated in Figure 6.7. The result are the three tristimulus values X, Y, and Z, where Y represents luminous reflectance, related to lightness. They uniquely identify a light stimulus reflected from the surface of an object with a given reflectance, as illuminated with a standard light source and as absorbed by the cones of the standard observer. If the stimulus is from a light only, the process is comparable, except that Y represents an open-ended luminance scale rather than luminous reflectance (see Chapter 4). In

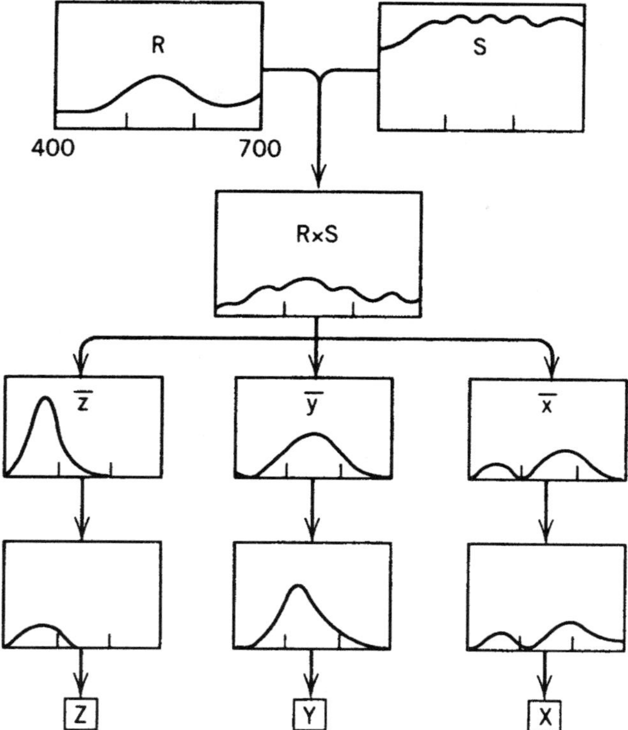

FIGURE 6.7 *Schematic representation of the calculation of the CIE tristimulus values* X, Y, *and* Z *for an object of reflectance R, as viewed by the CIE standard observer in light of a spectral power distribution S.*

practice these values are usually calculated automatically as part of measurement with a spectrophotometer, or the values are immediately used to calculate additional values defining the absorbed stimulus.

As mentioned, the three tristimulus value scales can be seen as axes (taken as orthogonal) of a space in which different color stimuli occupy unique locations. This is true in all cases except that of metamers. All spectral functions that are part of a given metameric suite (having identical tristimulus values) fall, when reduced to tristimulus values, on the same point in that space. As a result the space distinguishes between stimuli that require different amounts of three standard lights for matching. As is apparent, it does not indicate anything about the appearance of these stimuli. Geometrical distances between points in the space are unlikely to be in agreement with perceptual distances between color stimuli represented by these points and, as is shown in Chapter 7, they are not.

The shape of the color-matching functions dictates that color stimuli as absorbed by the cones cannot fill the tristimulus space completely. To demonstrate this it is instructive to display the locations of spectral stimuli in the space. For easier identification

(a)

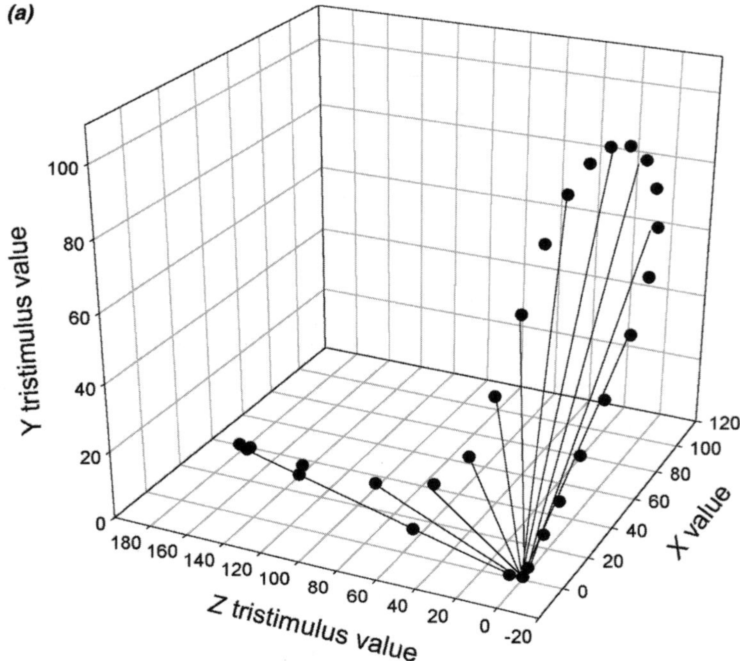

(b)

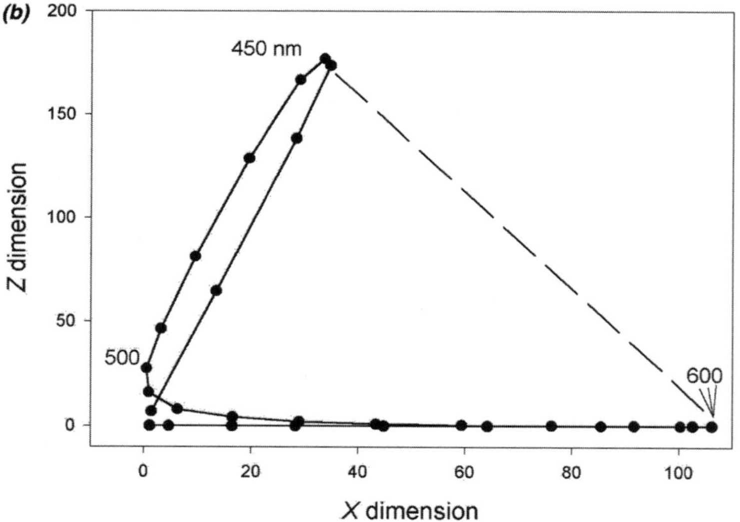

FIGURE 6.8 *(a) Vectors of spectral stimuli shown in the CIE X, Y, Z 2° tristimulus space in a three-dimensional view. For clarity only a few vectors are shown. (b) Endpoints of the vectors of part (a) shown in the X, Z plane of the 2° tristimulus space. The dashed line represents the location of red and purple colors not existing in the spectrum.*

the spectral stimuli are displayed in Figure 6.8a as vectors beginning at the origin of the space. These vectors form a butterfly-wing-like structure beginning and ending in zero (at the beginning and end of the visible spectrum). Since Y represents luminance, the brightness of each spectral light is immediately apparent. Spectral colors represent the chromatic limit in the space of color stimuli, and all object color stimuli must fall inside the boundaries in the X and Z dimension given by the spectral vectors. Figure 6.8b illustrates the vector endpoints in the X, Z plane of the space. Since the Y dimension is indicative of brightness or lightness, this plane must be indicative of the chromatic aspects of the stimuli, with Z roughly indicative of a yellowness to blueness scale and X of a greenness to redness scale (9). As mentioned, the spectrum, does not contain stimuli that are seen as red to bluish purple colors. These can be generated from mixtures of lights near both ends of the spectrum. These stimuli fall on the line connecting the ends of the two legs of spectral stimuli in Figure 6.8b, forming a triangle. Corresponding object colors have higher reflectance at both ends of the spectrum than in the middle.

By the mid-twentieth century it had become apparent that the spectral form of color-matching functions depends on the size of the visual field used in their determination. The CIE color-matching functions of 1931, as mentioned, apply to a 2° field of vision. Implications are that there is a change in average cone sensitivity, depending on field size (10). In 1964 the CIE promulgated a second standard observer, the CIE 1964 10° standard observer, applicable to a larger field of view. The corresponding color-matching functions are somewhat different and, as a result, so are related metamers and the chromaticity diagram. In practical work in color technology today, usually 10° observer data are used. But, as mentioned, the aim colors of the Munsell and NCS systems are defined in terms of the 2° observer.

CIE CHROMATICITY DIAGRAM

Given the difficulties of orientation in a three-dimensional space, the CIE elected to also recommend a chromaticity diagram containing information about color stimuli (as absorbed by standard cones) other than brightness/lightness information. It was recognized at the time that to be useful the information in the diagram must be normalized. As shown later in Figure 6.12, the solid of all possible object colors in X, Y, Z space has the form of a slanted spindle. The projection of the central axis of this form onto the X, Z plane forms a line, that is, achromatic colors do not fall on a point in the diagram, as is desirable. The normalization achieving this objective selected by the CIE divides individual tristimulus values by the sum of all three:

$$x = X/(X + Y + Z), \qquad y = Y/(X + Y + Z), \qquad z = Z/(X + Y + Z)$$

x, y, and z are called *CIE chromaticity coordinates*. The first two form the axes of the CIE chromaticity diagram (Fig. 6.9) (11). The horseshoe-shaped outline is the locus of the spectral stimuli. The straight line closing it off is the locus of the nonspectral purple stimuli. All possible color stimuli fall on or within the outline. In the interior of

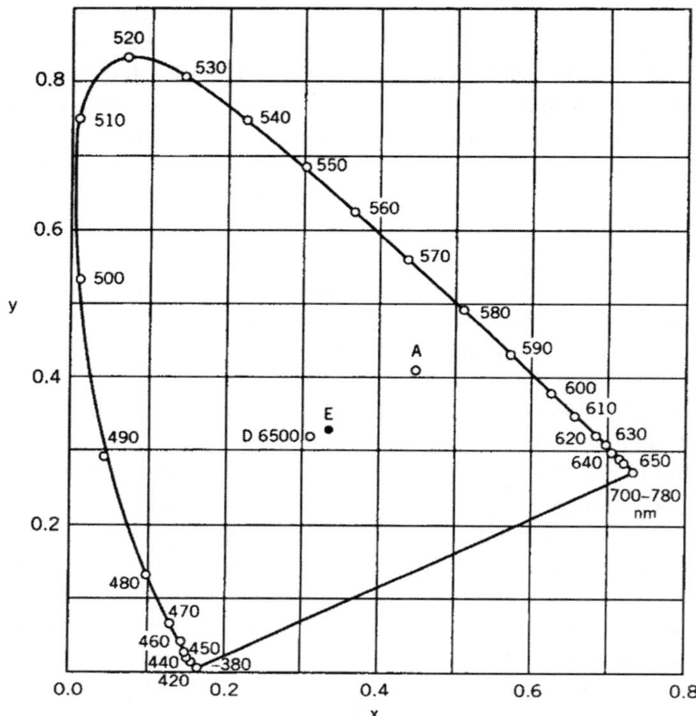

FIGURE 6.9 *CIE chromaticity diagram of the 2° standard observer. Spectral colors fall on the curved outline, purple colors on the solid line connecting the ends. The locations of the equal energy illuminant E, standard daylight D6500, and incandescent light A are also shown (8).*

the diagram, the positions of the equal energy illuminant (E) and of CIE illuminants D6500 and A are shown. It has been and continues to be tempting to print this figure filled with pigment color stimuli. This is misleading in several respects.

The chromaticity diagram is linear in nature and has the advantage that the results of the additive color mixture (mixture of lights) can be represented in simple manner. For example, the results of mixing lights of 470 and 575 nm fall on a straight line connecting the two (and passing through the equal-energy point), forming a desaturation line between blue- and yellow-appearing spectral colors. On the other hand, mixtures between 550- and 650-nm lights (appearing yellowish green to yellowish red) fall on a straight line represented by the spectral stimuli in between. An appropriate mixture of the two has the pure yellow appearance of the corresponding spectral light. Such procedures are only applicable in the case of lights, but not in the case of object color stimuli.

The relationship between locations in the chromaticity diagram and average color perceptions is complex, as will become apparent in Chapter 7. As rough general indicators, it has been the practice to identify hue with wavelength or complementary wavelength (for purple colors, wavelength at the other end of a line passing from the purple color locus through the illuminant point). A measure of saturation called *purity* has been defined as the ratio between the length of the line from the central illuminant point to the locus of the color stimulus inside the diagram compared to the locus of the spectral color on the extended line. But these are of little practical value. The basic CIE colorimetric system does not include normalization in the form of an opponent color system. This is only part of color difference or color appearance models. Balanced linear opponent color functions can be calculated by subtracting the $\bar{y}$ from the $\bar{x}$ and the $\bar{z}$ from the $\bar{y}$ functions and weighting the results appropriately for equal loop weight.

As mentioned, the chromaticity diagram does not contain brightness/lightness information. This information has been added in a specification system consisting of x, y, Y, that is, the two chromaticity coordinates and the tristimulus value Y representing luminance or luminous reflectance. This space has little practical value because of its lack of perceptual uniformity and the failure of the additivity law (see Note 4 in this chapter) in certain conditions. Historically, it has been used to demonstrate the shape of the optimal object color solid.

OPTIMAL OBJECT COLOR (STIMULUS) SOLID

Reflectance functions of objects have to fit into an area defined by the axes visible wavelength (from about 400 nm to about 700 nm) and reflectance factor (from 0 to 1). Within this diagram there are an indefinite number of possible reflectance functions. Many of these represent functions metameric to others when viewed in a given light by one of the standard observers. Real objects usually have reflectance across the whole visible spectrum. When viewed in broadband lights, such as daylight, the stimulus arriving at the eye is also broadband. An interesting question arises as to what are "optimal" object color stimuli. Optimal in this respect means having the highest possible chroma at a given level of lightness. This question was addressed in the early twentieth century by Ostwald, and soon after, in a more fundamental way, by Schrödinger (12). According to Ostwald and Schrödinger, there are two basic types of optimal object color reflectance functions. They and their mirror functions are illustrated in Figure 6.10. They have sharp transitions at given wavelengths, in one case, a single one, in the other case, two. Stimuli change as a result of the transition wavelengths. By traveling through all possible transition wavelengths and by adjusting the height of the functions to represent given values of luminous reflectance, it is possible to construct a surface of optimal color stimuli in the CIE tristimulus space or the x, y, Y space. All real object color stimuli have to fall within the surface of the solid (13).

Shortly after the CIE colorimetric system had been established, such solids were calculated by MacAdam for optimal object colors viewed by the standard observer in daylight illuminant C (an early version) and in tungsten illuminant A. The shape of the solid differs as a function of the illuminant. The result for illuminant D65, calculated later, is shown in Figure 6.11. All possible object color stimuli for this illuminant fall on or within the surface. The central axis represents achromatic colors

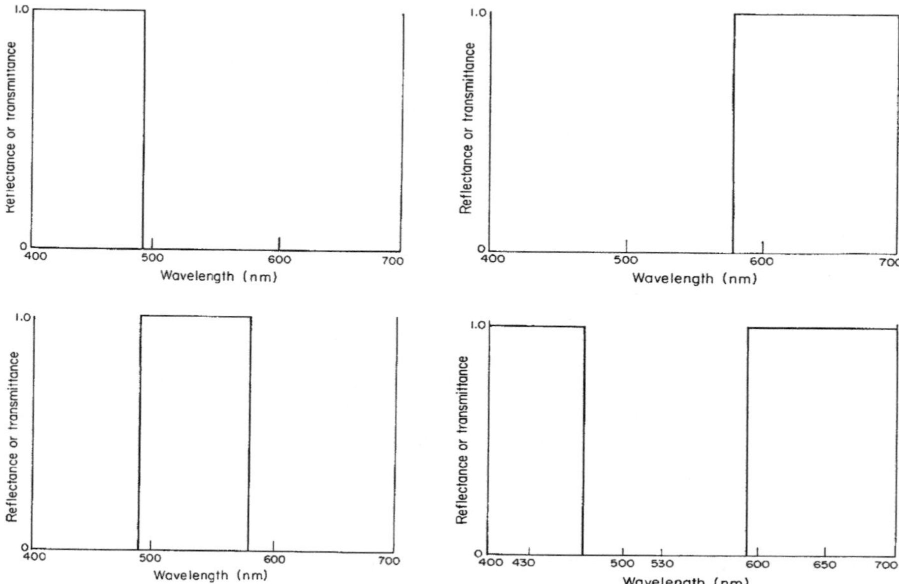

FIGURE 6.10 *Types of optimal object reflectance functions with one and two transitions.*

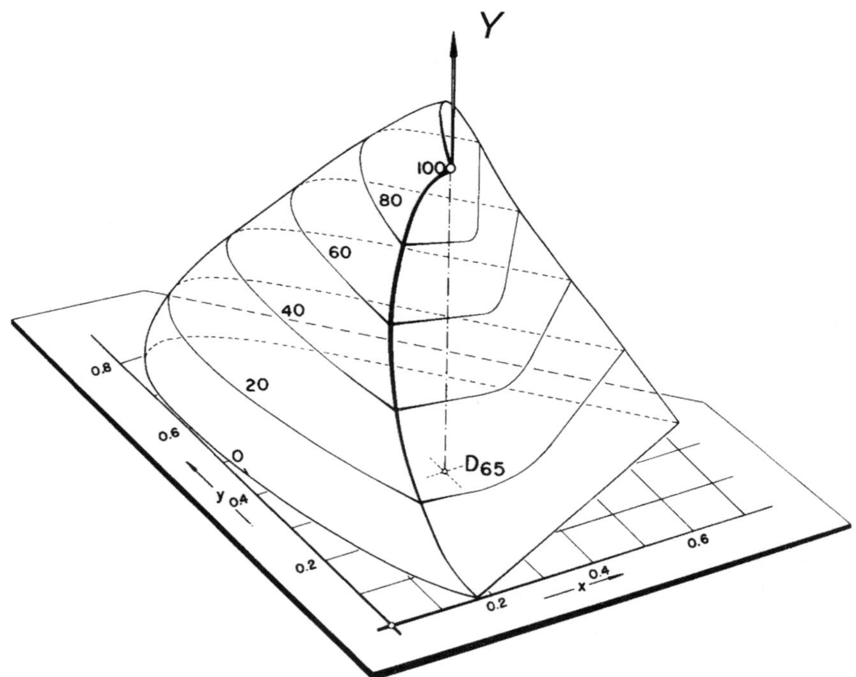

FIGURE 6.11 *The optimal object color solid for illuminant D6500 in the CIE* x, y, Y *space (8).*

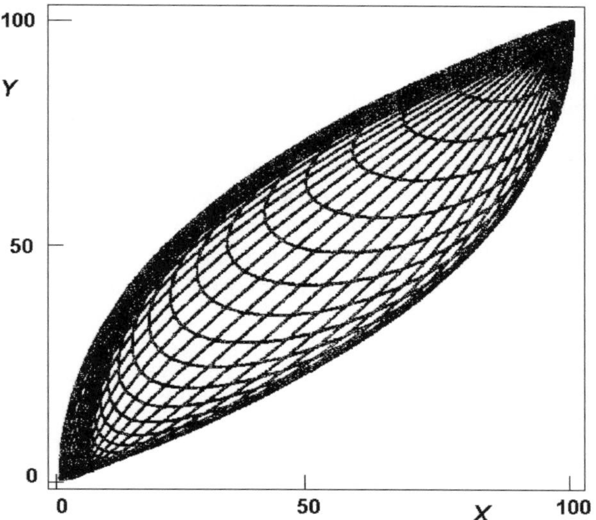

FIGURE 6.12 *View of the optimal object color solid in the* X, Y *plane of the CIE tristimulus space (14).*

with zero luminous reflectance on the plane and luminous reflectance 100 on top. While instructive of the geometrical realities of the object color stimulus solid, this space, like the diagram it is based on, has no relationship of practical value to color appearance. An illustration of the optimal object color solid in CIE tristimulus space is shown in Figure 6.12.

This chapter presented the system used internationally for defining color stimuli in the form of lights or of light reflected from objects. The system is based on determination of spectral cone sensitivity functions using the technique of metamerically matching the appearance of spectral stimuli using three lights widely separated in the spectrum. In the complete colorimetric system, lights are defined in a relative manner by their spectral power distributions and objects by their spectral reflectance functions. The system is effective for its original purpose. In practice light sources are rarely in close agreement with one of the CIE-specified illuminants and the intensity of the light is not considered. Reflectance functions depend on the specific instrument geometry implemented in a commercial instrument. Color-matching functions vary by observer (to the extent that perhaps 90% of all color-normal observers have color-matching functions different in such a manner that many of one observer's matches are seen as mismatches by another) as well as by field of view. They represent the very simple perceptual situation of looking at a bipartite field against (usually) a black surround. These facts indicate that there are a considerable number of variables that can affect the relationship between actual color stimulus and the signal leaving the cones. Most importantly, however, the colorimetric system does not take any account of all the processing between the cones and the final color experience. As was mentioned and will be shown in more detail in the next chapter, the colorimetric system is used as a basis for models that attempt to predict color experiences by an average observer (as represented by the standard observer). The success is less than perfect, but substantial. Barring a fuller understanding of the perceptual system, it is as good as can be achieved based on the historical record of experimental data (itself less than perfect, as indicated in Chapters 4 and 5).

One of the commercially most important areas of color technology is the control of color in manufacturing of colorants and colored objects. This will be discussed in more detail in Chapters 7 and 8. Another is the rapid, accurate formulation of colorant recipes for production of goods matching color standards. This is the subject of Chapters 8 and 9.

7

Calculating Color

The idea of calculating taste or smell may be considered far-fetched, but calculation of color (or at least some aspect of it) is taking place every day. There are several reasons making this possible, two of which are (1) color stimuli from uniform fields can be physically measured with good accuracy, and (2) the fact that there are only three relatively well-defined types of light sensors, the cones, responsible for providing input to our color vision system seems to restrict the problem to three dimensions, something humans can comprehend without much difficulty. The realities faced by a color vision model attempting to predict anything related to perceptual judgments of color are much more complex, however, as builders of such models know. The reason is that the human color vision system has developed for the purpose of solving problems critical for our survival by modifying the signals existing at the end of the cones in many different ways, depending on many factors. The final results may be of a Bayesian statistical nature. Any model short of duplicating in a robotic fashion the complete color vision system will be found inadequate in specific circumstances.

Mathematical models of a complex process can be built in two different ways: (1) by understanding the process completely and building the model accordingly, and (2) by understanding a few aspects and using these and statistical information about regularities in the data to be modeled that are not predicted by the basic model and finding mathematical fixes to take account of them. In the absence of a complete understanding of the system, the second method is what has been pursued over the last 150 years. Given the complexity of human vision, there is no model at this time

Color: *An Introduction to Practice and Principles, Second Edition*, by Rolf G. Kuehni
ISBN 0471-66006-X Copyright © 2005 John Wiley & Sons, Inc.

that can predict with reasonable accuracy a wide range of phenomena. Models usually are fine-tuned to predict a limited number of data, and certain parameters in the model need to be changed to predict others.

It is a truism that models can only be as good as the data they are to represent. As discussed in Chapters 4 and 5, data we have on perceptual color judgments of, say, which stimuli represents unique hues, which pairs of stimuli represent perceived differences of equal magnitude (or of equal hue, chroma, or lightness), which object color stimuli are seen as approximately the same in different lights, which stimuli are metameric, and so on, vary considerably by observer. As a result, predictions can only apply with accuracy to a defined standard observer and the lucky few (perhaps 10% of the population) whose color vision system is in good agreement with the mean. It is already very clear that perceptual results are not primarily dependent on an individual's specific cone functions, but that many other neural manipulations of those data can lead to many different results in specific situations.

Given that difficult set of starting conditions, it is not surprising that building models for certain aspects of color vision has been an arduous task and that the results, from the perspective of an individual's system, are mixed. A discussion of such efforts in this chapter is limited to modeling of color space and its division for a specific set of conditions: color difference calculation. Additional types of calculation will be briefly discussed in Chapters 8 and 9.

If we assume, as has been done in Chapter 5, that all possible color perceptions can fit into a three-dimensional space (itself dependent on what distances in the space they are to represent), then the organization of that space can be viewed from two perspectives: (1) a global one that establishes the form of, say, the object color solid in the space and then finds the way in which it is to be scaled internally to be in agreement with perceptual results, and (2) a bottom-up approach in which the global space is built by assembling just noticeable differences from a starting point, say, a middle gray. It might be expected that both approaches mesh in some way. Knowing already about effects such as lightness and chromatic crispening and the HKE provides an idea of the complexities the model builder faces. As Chapter 4 has shown, color perceptions from stimuli vary as a function of surround lightness and chromaticness. This issue was initially disregarded and only in more recent years has it received more recognition, of necessity, by builders of more complex models such as color-appearance models. In its logical conclusion, this means that a model can only be reasonably accurate for a very specific set of conditions of stimulus presentation, surround, illumination, and questions asked of the observer. Changes in these conditions require changes in the model.

For a number of reasons modeling of global space or by small differences has been limited to uniform or isotropic organization, a structure where geometrical distances of equal size in any direction correspond to perceptual distances of equal magnitude, such as those attempted, with more or less thoroughness, in the Munsell and OSA-UCS systems. There is no known attempt to mathematically model the Hering-type space, such as represented by NCS, perhaps because there is no practical application with a need for it.

MODELING GLOBAL COLOR SPACE

Until very recently there has been no specific attempt to build an accurate mathematical model of the Munsell system based on tristimulus data of its aim colors. Its limitations have been recognized since the 1950s and the work performed in connection with developing OSA-UCS. The development of the latter has shown that a Euclidean model of isotropic global color space is impossible. Instead it requires a space with a surface of a curvature that cannot be geometrically represented in Euclidean space, due to the effect of hue superimportance (Chapter 4).

Before this fact was known Adams developed in 1942 a simple mathematical model of color vision that became the leading idea in the technological development of color space and color difference formulas (1). The key idea behind the model expresses a psychophysical fact known long before: the relationship between tristimulus values and the perceived magnitude of steps between stimuli is not linear. The larger the magnitude of the stimulus in terms of tristimulus value, the larger the change in stimulus required to perceive a step of standard size (not considering crispening effects). The reason for this compression of stimulus is believed to lie in saturation effects in the cones. The more light quanta they absorb, the less well they can respond to them. In the mid-nineteenth century the experimental psychologist Fechner proposed that the relationship (not just in case of vision) is geometrical, represented by a logarithmic scale. Detailed experimental work has shown that in the case of color the relationship depends on several conditions and is usually less than geometrical. According to Stevens it can be represented by a power scale (2). Figure 7.1 illustrates the relationship between luminous reflectance and perceived

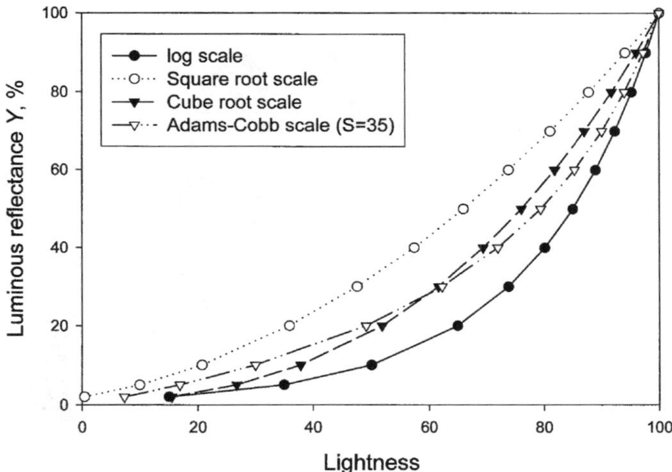

FIGURE 7.1 Relationship between luminous reflectance and metric lightness (not considering lightness crispening and the HKE) for different compression models: geometrical (log scale), square and cube root, and the Adams–Cobb scale of 1922 (surround luminous reflectance Y = 35).

lightness according to four stimulus-compression models (lightness crispening and HKEs are not considered, however). When the experimentally determined Munsell lightness scale was measured, smoothed, and averaged during development of the renotation aim colors, a complex polynomial formula was fitted to it. Later it was shown that the polynomial is well fitted with a cube root, that is, Munsell value as defined changes as the cube root of luminous reflectance (3). Adam's idea was to apply the Munsell lightness polynomial not just to luminous reflectance as expressed with the Y tristimulus value but to all three tristimulus values. To achieve normalization of the system (see Chapter 6) he subtracted the resulting values for Y from those of X and Z, thus obtaining an opponent color space model. Adams also proposed that different weights be applied to the two opponent color signals to bring them into better agreement with perceptual results. In 1976 the CIE promulgated the CIELAB color space and difference formula, a cube-root version of the Adams space (4). At the same time, it also recommended the CIELUV formula. The chromatic diagram

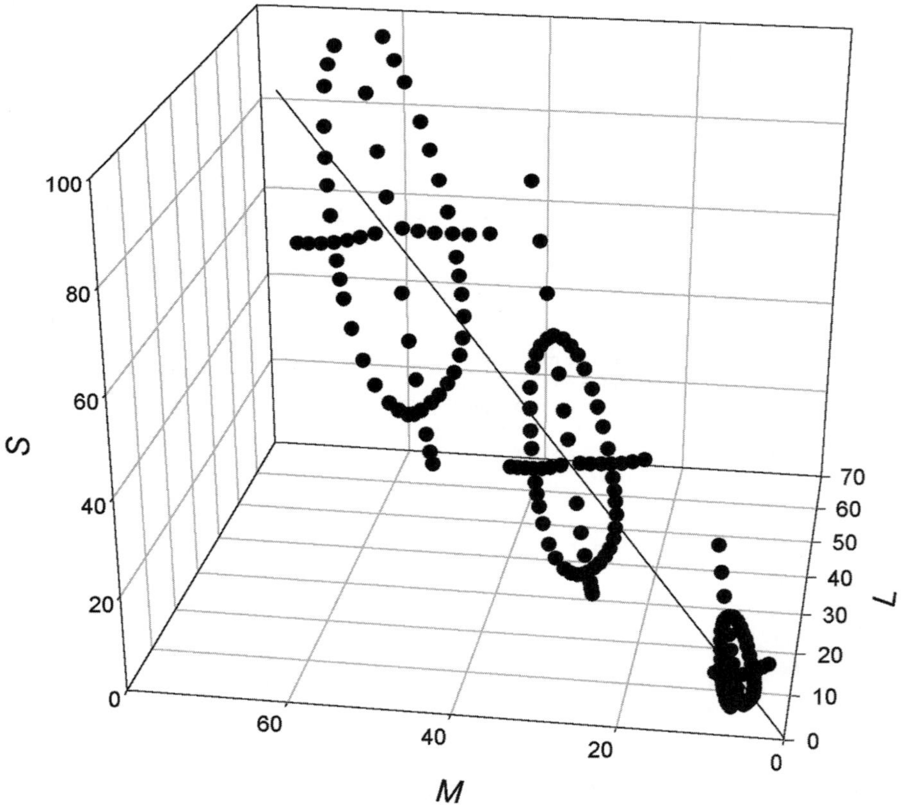

FIGURE 7.2 *Three Celtic crosses (see text for details) of Munsell aim colors plotted in the* L, M, S *cone sensitivity space.*

of this formula is a linear transformation of the CIE chromaticity diagram, modified from a formula proposed in the 1930s by Judd (5). As a linear transformation it does not have signal compression, except in the lightness formula that is identical to that of the CIELAB formula. It has not found usage for object colors.

To demonstrate the steps from cone space to CIELAB space, three Celtic crosses of Munsell color stimuli are used. They consist of 40 hue circles at chroma 8 and values 3, 6, and 8, as well as additional chroma steps for hues that fall on or near the chromatic axes of CIELAB space. In Figure 7.2 they are illustrated in the L, M, S cone sensitivity space. Here they have elongated contours that are angular slices through an elongated cone centered on the equal energy line originating at the origin of the diagram. Chroma scale colors fall on curved lines and are spaced irregularly. However, the three value levels are well separated from each other, and hue and chroma steps are in ordinal order compared to the perceptual organization. The question arises how to modify the model so that it is in better agreement with that organization. A first step is to linearly transform the reference frame from L, M, S to the CIE tristimulus values X, Y, Z where one dimension is in general agreement with perceived lightness. This step is illustrated in Figure 7.3. There continues to be an elongated cone originating at the zero point, but the crosses now form horizontal slices in it. The contours are still strongly elliptical and the chroma scales irregular. The three-value-step difference between values 3 and 6 appears to be of comparable magnitude to

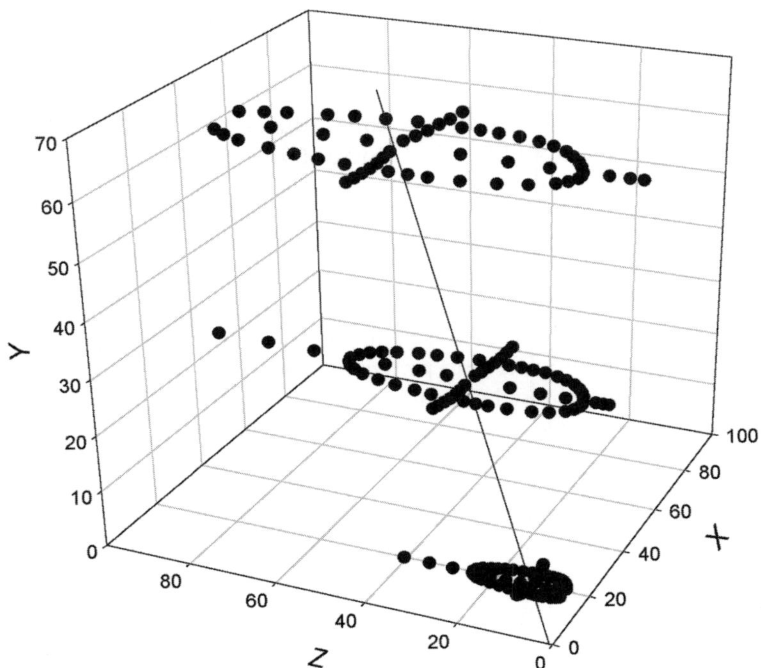

FIGURE 7.3 *The Celtic crosses of Figure 7.2 in the CIE 2° X, Y, Z tristimulus space.*

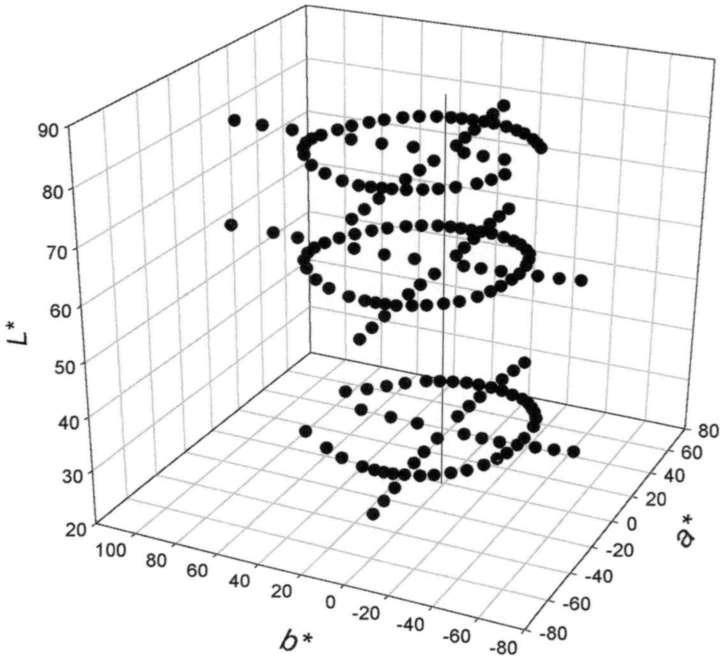

FIGURE 7.4 *The Celtic crosses of Figure 7.2 in the CIELAB color space.*

the two-step difference between 6 and 8. Ellipses can be converted to near circles by applying a weighting factor (2.5 in the CIELAB formula) to the X scale. The system requires normalization so that the cone is upright and is converted to a cylinder. The former is achieved by normalization by subtraction of tristimulus values, the latter by application of power compression. The result is shown in Figure 7.4 and indicates that the contours are more circular, the cone is righted and has become a cylinder, and the value levels are in better agreement with perceptual results. However, the result is still less than perfect, as shown in Figure 7.5. Ideally, all three contours should be identical in this figure and form perfect circles. Given the fact that there is no reason to consider the Munsell system perfect, we should not be surprised. On the other hand, as indicated, we cannot assume that the model is a good representation of idealized human color vision. The model does not take account of the relative size of the Munsell hue and chroma differences, and therefore can fit into a Euclidean space.

Figure 7.6 provides an idea of how well a Euclidean mathematical model of the OSA-UCS experimental data predict those data. The 43 samples used in the experiment are plotted in the chromatic diagram fitted to the results. If the diagram were perfect, the perceptual distances between the points would all be the same. Arrows and bars give an indication of the average magnitude of the perceptual differences. Where the magnitude is smaller than expressed in the formula, the gap is indicated by the space between arrow points. Where it is bigger, the missing space is shown by the

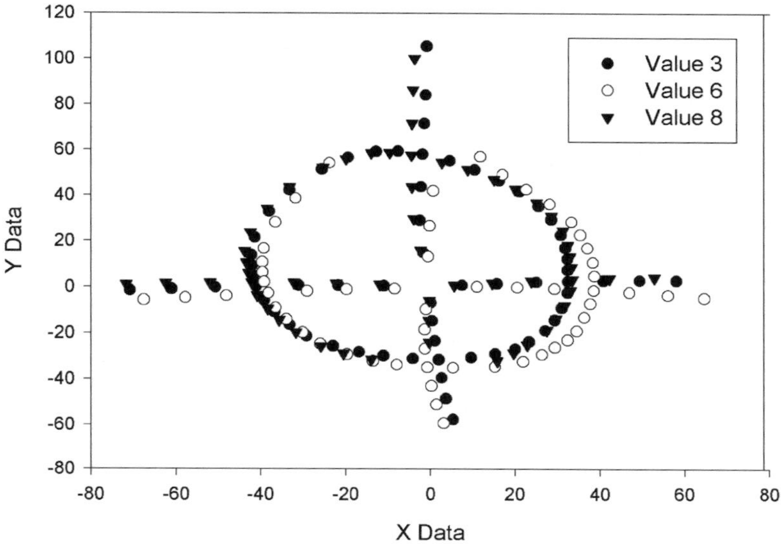

FIGURE 7.5 *Projection of the Celtic crosses of Figure 7.4 onto the* a*, b* *chromatic plane.*

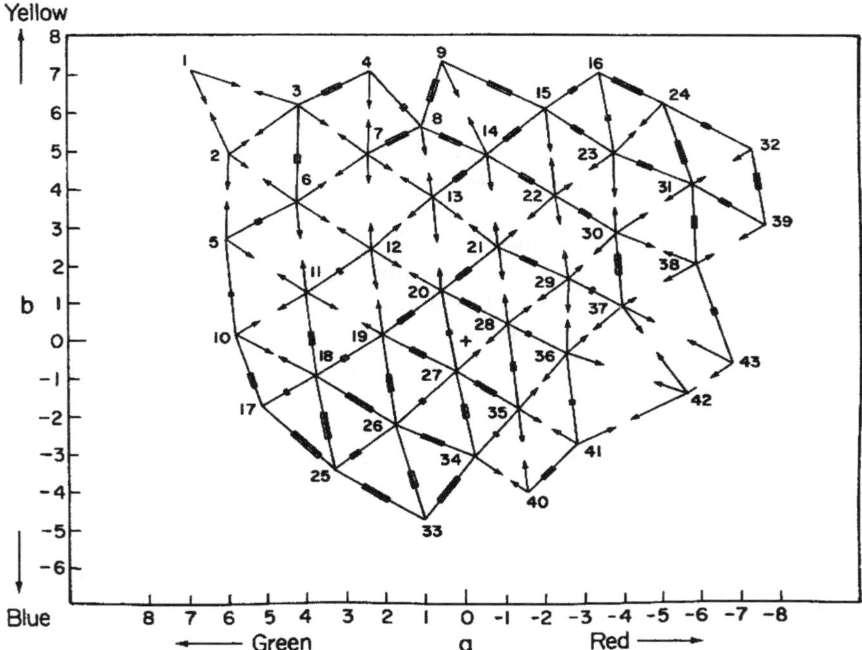

FIGURE 7.6 *Plot of 102 chromatic differences between 43 color samples in the chromatic diagram of the Committee on Uniform Color Scales (6). Numbers identify the samples. Lines with arrows and bars show the size of average visual difference judgments. The colorimetric difference is too large in the case of arrows and too small in the case of bars.*

length of the bar placed over the connecting line. The results show a surprising degree of difference between an optimal fitted Euclidean model and perceptual results, in part due to the hue superimportance effect. The agreement between model and data is significantly improved by introduction of a hue superimportance factor.

The results of modeling of the average perceived differences of the magnitude represented in the Munsell and OSA-UCS systems do not inspire confidence that such models perform well for color difference data at the just noticeable difference level. But for reasons of uniformity of usage and convenience modifications of CIELAB have been used exclusively for such efforts since its inception.

SMALL COLOR DIFFERENCES

The smallest perceivable color differences are the just noticeable differences (JND). In the late nineteenth century, Helmholtz assumed that Fechner's law was directly applicable to the three color fundamentals (the cone responses) and that it applied equally to all three (7). He proposed a so-called line element, in which he calculated the difference between two stimuli as the square root of the sum of the squares of the differences in the three responses appropriately compressed, that is, he assumed color space to be Euclidean. Comparison against perceptual data quickly showed that this approach was too simple. In succeeding years more and more complex line elements were proposed, but they found little practical application. In 1942 MacAdam published an empirical line element (8). His assumption was that the magnitude of a JND was guided by the color-matching error. In his experiments one observer repeatedly made metameric matches of constant brightness with three lights against 25 reference stimuli distributed roughly evenly across the chromaticity diagram, using a visual colorimeter. The matching error for each reference was treated statistically and fitted with a unit error ellipse around the standard. Figure 7.7 is an illustration of the resulting ellipses (ten times enlarged). Subsequently several mathematical formulas were developed based on these ellipses that allowed calculation of small color differences. When they were tested against perceptual data involving object colors, they were found to perform inferiorly to other formulas, for example, the one based on the Adams's model. The reason for the failure of the MacAdam ellipses for purposes of small color difference calculation was only recently discovered: color-matching error appears to be due only to the relative sensitivity of the cone types and does not involve the separate mechanisms that seem to be responsible for hue and chroma difference evaluation (see Chapter 4) (9).

In the 1950s and 1960s interest in the calculation of small color differences for purposes of quality control increased because of improved calculation capabilities and the availability of relatively inexpensive equipment for the measurement of reflectance. In the same period, several color difference experiments using textile and painted paper samples were performed, offering perceptual data for testing of formulas. In the mid-1970s, more than a dozen different color difference formulas were in industrial use, producing growing confusion because the results from different

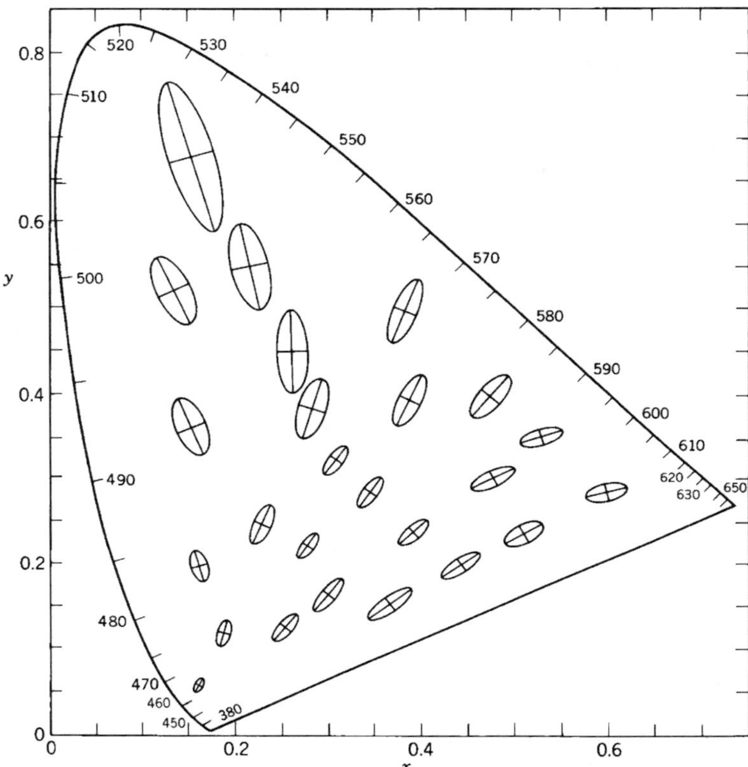

FIGURE 7.7 *Ellipses fitted by MacAdam to color-matching error data for one observer around 25 standard lights, ellipses 10 times enlarged, CIE 2° chromaticity diagram (8).*

formulas could not be compared easily. As a result, the CIE proposed the CIELAB and CIELUV formulas in 1976, not because they performed better than other formulas, but for "uniformity of usage." The optimal object color solid of CIELAB is illustrated in Figure 7.8. The axes of its space are L^* for metric lightness, a^* for, in a general sense (not to be confused with unique hues), greenness–redness, and b^* for, in the same general sense, yellowness–blueness. The CIELAB difference formula has the advantage that it can be expressed in terms of both rectangular as well as polar coordinates. In the latter form, the coordinates are metric lightness L^*, hue angle h_{ab}, and metric chroma C^*, that is, an arrangement in terms of Munsell attributes. This form has significant advantages because of the matter of the hue superimportance. The two forms are schematically illustrated in Figure 7.9.

In the industries dealing with colored materials, the performance of the CIELAB color difference formula was immediately considered unsatisfactory, and efforts were begun to improve it by studying regularities in the perceptual data. Fitting unit contours

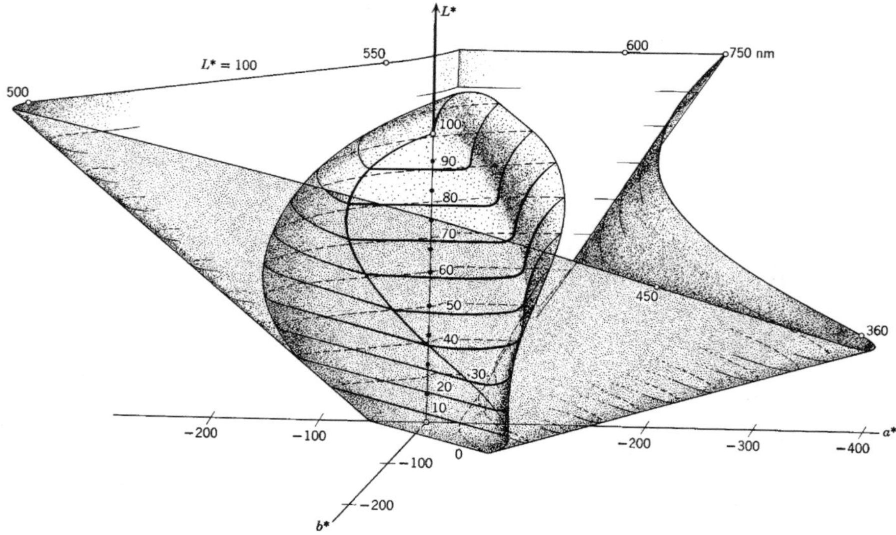

FIGURE 7.8 *The optimal object color solid in the CIELAB* L*, a*, b* *color space (10). The envelope of spectral colors with a curved purple line (due to nonlinear compression) is also shown.*

in form of ellipsoids to various sets of small color difference data resulted in the following insights (applicable to a gray surround):

- In the chromatic plane unit contours are elongated ellipses pointing, in general, in the direction of the neutral point of the diagram (see Figure 7.10), as a result of hue superimportance.
- Ellipses (and ellipsoids) increase in size as a function of metric chroma, the result of the chromatic crispening effect.

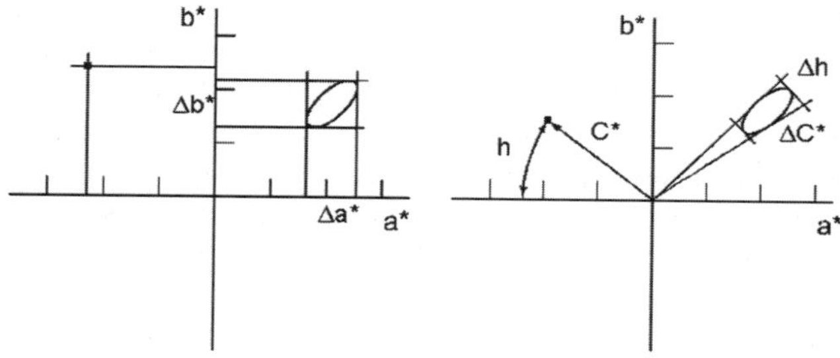

FIGURE 7.9 *Schematic sketch showing the comparison between rectangular and polar coordinate versions of the CIELAB chromatic diagram. The ellipse illustrates that the latter version is better able to deal with such unit contours.*

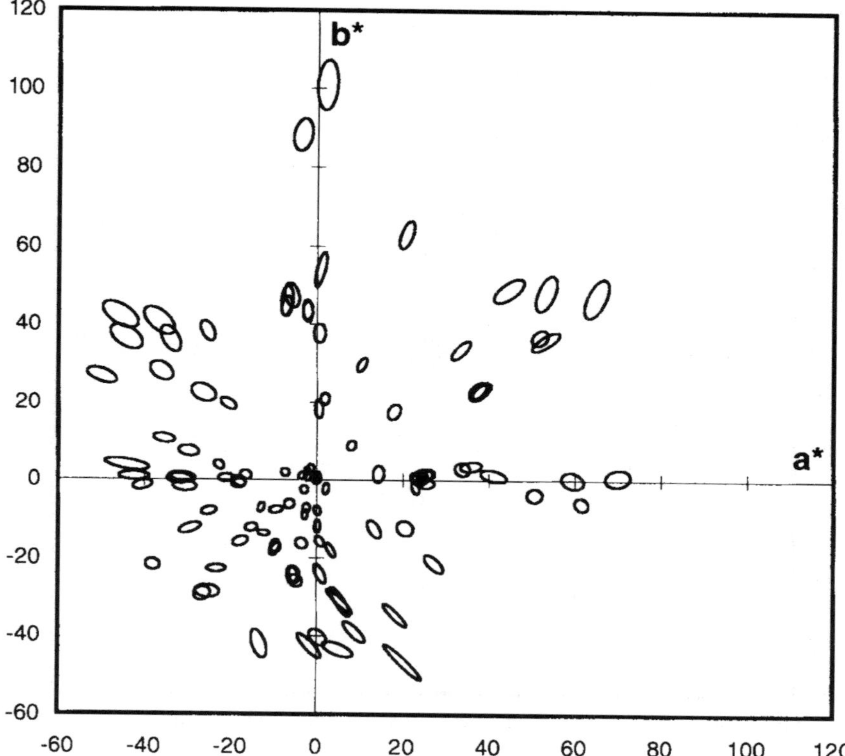

FIGURE 7.10 *Ellipses fitted to the normalized set of BFD perceptual small color difference data in the* a*, b* *chromatic diagram (11).*

- Ellipsoids increase in size in the lightness direction the farther away they are located from the metric lightness of the surround, the result of lightness crispening (12).
- Hue differences are not linearly related to hue angle differences, but rather in a complex, nonlinear fashion.
- Ellipses near the negative b^* axis are tilted in a counterclockwise direction, perhaps as the somewhat accidental result of the definition of the $\bar{x}$ color-matching function (13).
- Unit contours near the neutral point continue to be elongated.

These issues (more or less explicitly and more or less completely) have been addressed in subsequent versions of color difference formulas, such as the Color Measurement Committee (CMC) formula developed in England and widely used in industry, and the latest recommendation by the CIE, named CIEDE2000 (11). Figure 7.10 illustrates unit difference ellipses fitted to various sets of normalized

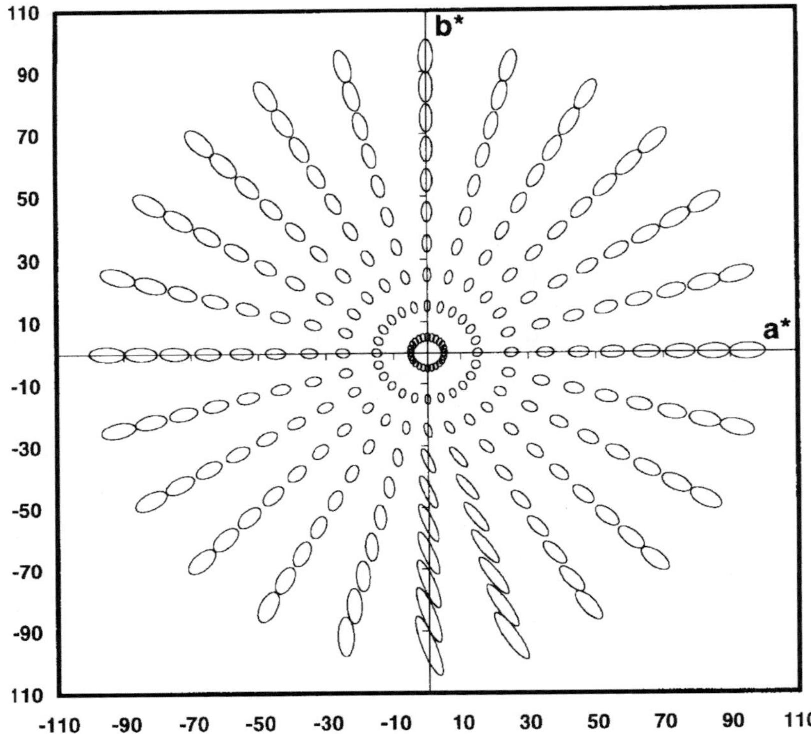

FIGURE 7.11 *Unit ellipses in the* a*, b* *diagram calculated from the CIEDE2000 color difference formula (13).*

perceptual data in the a^*, b^* chromatic diagram. For comparison, Figure 7.11 shows, the unit ellipses in the same diagram converted by the CIEDE2000 formula to circles of equal size. There are additional regularities in the data, so far not addressed, whose effect on accuracy of prediction is not yet known.

Improved formulas of this type are based on the foundation of CIELAB, that is, they modify the CIELAB frame locally. In this way they make treatment of the non-Euclidean nature of uniform color space possible within the Euclidean geometry of CIELAB. Compared to CIELAB, with an average (for different data sets) accuracy of predicting average perceived color difference of about 50%, CIEDE2000 predicts it with approximately 65% accuracy.

The reasons behind this modest level are as yet unknown. The level of accuracy is very likely the combined result of variables involving observer panels as well as variables involving experimental conditions. Replication experiments of perceptual data have as yet not been performed. It is not known if this level of accuracy represents the best that can be achieved or if by controlling experimental conditions more tightly than has been done in past experiments accuracy can be improved. The price for higher accuracy may be having very restricted experimental conditions, with less and less true applicability to conditions in the real world.

Despite these limitations, colorimetric color control, as the only more or less objective game in town, is used widely in quality control both within as well as between companies. But, given limitations of reflectance measurement for many structurally complex materials or materials with complex coloration (woven fabrics, prints, etc.) and contrast effects affecting esthetics, final judgment is still often made by a human arbiter.

Because of the fading of the chromatic crispening effect as the size of differences increases, color difference formulas such as CIEDE2000 only have the cited level of accuracy in a limited range of small differences, from zero to about five units. Beyond that they become increasingly less accurate.

The difficulties of obtaining accurate predictions of perceived differences even for very limited conditions of viewing provides an idea of the difficulties of accurately predicting color appearance in more complex, more natural situations. Given the potentially large variability in human color perception, not completely understood at this time and made possible by our innate capabilities to adapt to given situations, it is not likely that relatively simple models can be very meaningful for an average observer. There is also the question of how much more complex models have to become (that is, how much closer they have to match the complexity of the total visual apparatus) before close predictions become possible. Perhaps sometime in the future it will be necessary to merely set a few personal parameters in a complex model to obtain accurate predictions according to one's personal experience.

8

Colorants and Their Mixture

Colorants are materials having absorbing and, in the case of pigments, scattering properties. Scattering of a colorant depends on the size of its molecules, its solubility in the substrate, and its tendency to form crystals. Colorants with small molecular size and good solubility are usually dissolved monomolecularly in the substrate, that is, each colorant molecule is separated from the other. Depending on solubility and attraction forces, some colorants form aggregates in the substrate. If the solubility is poor, colorants precipitate, usually in crystalline form. In that case there continues to be absorption but, depending on the particle size, there is also more or less scattering of light on the crystal surfaces. Light striking colorants is absorbed in a spectrally selective way due to the behavior of certain electrons in molecules (see Chapter 1). Technically, the distinction between dyes and pigments is somewhat ambiguous. Vat dyes are called dyes even though in their final form on the textile substrate they are in the form of crystalline particles that also scatter light. Pigment dispersions are also used to 'dye' textile fabrics or paper. Pigment lakes can be formed from dyes by precipitating them with chemicals that have affinity to them and that, in combination, form amorphous or crystalline powders. Regular pigments have a chemical structure, making them poorly soluble in most solvents and forming crystalline particles. They are finely ground to optimize absorption and scattering properties. However, pigments are distinguished by their degree of transparency, that is, the crystals they form are more or less transparent depending on impurities and how the crystals fracture in the grinding process. Dyes in transparent media result in volume color experiences, pigments result in translucent or opaque materials. In fact, pigments are used to make transparent or translucent media opaque, for example, synthetic fibers or paint media.

Color: *An Introduction to Practice and Principles, Second Edition*, by Rolf G. Kuehni
ISBN 0471-66006-X Copyright © 2005 John Wiley & Sons, Inc.

Despite the fact that a paint medium may be transparent, the paint is opaque if it contains scattering pigment in sufficient quantity so that there is no longer transparency.

DYES

Dyes having absorbing properties only can be used to color transparent materials such as films, liquids, and plastics. For technical reasons they are also used to color many translucent materials, such as textiles and paper. They can also be used to impart surface coloration to opaque materials without affecting the surface structure, such as in the case of leather, dyed aluminum, or stained wood.

If a dye is dissolved in a transparent medium, say water, the depth of coloration of the water depends on the intrinsic coloring power of the dye molecules and on their concentration in the water. Up to a point, an increase in dye concentration will result in an increase in chromaticness of the color experience obtained when a light beam passes through the solution. The experience of color is due to the partial spectral absorption of light by the dye. The absorption effect can be measured by comparing beams passing through the solvent alone to those passing through the solvent containing dye. Different kinds of dyes cause absorption in different regions of the spectrum. The absorption bands can be relatively narrow or broad. Figure 8.1 illustrates the spectral absorption functions of narrow-band and broadband "violet" dyes. Such curves tend to be rather specific concerning the molecular structure of the dyes involved and can be used as fingerprints in dye identification. When dyes are dissolved in monomolecular form light absorption varies in relationship with dye concentration. But absorption is not linearly related to light transmission: the relationship is logarithmic. Absorbance

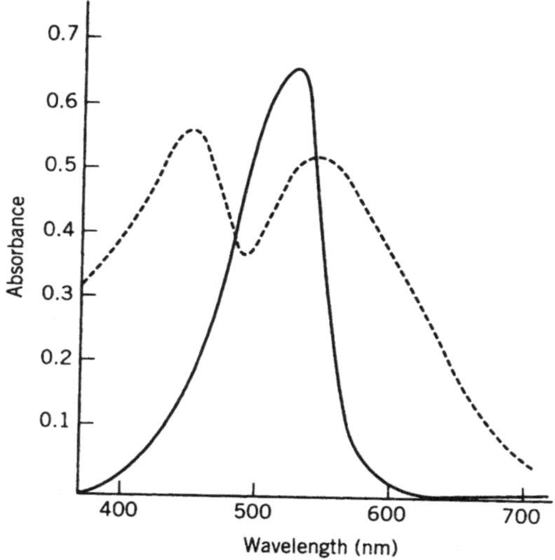

FIGURE 8.1 *Absorbance curves of narrow-band absorbing (full line) and broadband absorbing (dashed line) dyes dissolved in water.*

is defined as the logarithm of the inverse of transmittance ($1/T$) and ranges from 0 at 100% transmittance to 3.0 at 0.1% transmittance. The relationship between transmittance and absorbance is known as the Lambert–Beer law (1).

In principle absorbance values are additive. The absorbance of a mix of dyes, each at a given concentration, is ideally the sum of the absorbances of the individual dyes. Because of the proportionality of dye concentration and absorbance the absorption functions of various mixes of dyes can be predicted in principle from the absorbances of the dyes in unit concentration and their relative concentrations. In the reverse, unknown concentrations of dyes can be determined by measuring their transmittance and comparing the related absorbance values to those of a standard solution of the dye. Within limits it is possible to do this in complex mixtures of different dyes. In practice there are many limitations due to agglomeration of molecules in solutions of single dyes or the interaction of different dyes in a mixture. Analysis of dye concentration by transmittance measurement is used extensively in their quality control, however.

Dyes do not reflect light by themselves, but absorb light on the way to the opaque material to which they are applied. This material usually has some absorption properties of its own (think of leather or wood). The measured reflectance is thus a function of the absorption of the dye and the substrate as well as of the reflection or scattering properties of the substrate. For comparisons related to a single substrate, its reflectance/scattering can be treated as having the value 1 and thereby neglected. Accurate comparisons between different substrates require consideration of the reflectance/scattering properties of the substrate materials.

Dyes can be of natural origin or synthesized. Among natural dyes already known in antiquity are indigo, purple, madder, kermes, cochineal, gamboge, saffron (2). The synthetic dye industry had its start in the middle of the nineteenth century with the accidental synthesis of the dye mauvein by the British chemist Perkin (3). As mentioned in Chapter 1, since then tens of thousands of dye molecules have been synthesized, a few thousands achieving commercial success at different times.

Dyes with special properties are required for use in color photography. Three different layers in the film contain silver halide and a sensitizer that is active in a specific spectral region. After exposure that activates silver ions in the layers, the dyes are created during the development process. Treatments with a developer and three different couplers result in formation of yellow, magenta, and cyan dyes in the respective layers (in positive-image film). As a result of subtractive color mixture a large range of color stimuli can be obtained when light is transmitted through the developed film and the result viewed directly or reflected from a white screen. Typical transmittance curves of such dyes are shown in Figure 8.2, the yellow dye being CC, the magenta dye BF, and the cyan dye AG. Figure 8.3 illustrates the so-called chromatic gamut, the range of chromaticities that can be obtained from a mixture of the three dyes in different concentrations when viewed by the standard observer in daylight illumination. The relationship between changes in dye concentration and resulting chromaticity is nonlinear in a complex manner. A very sophisticated process of dye development is used in dye transfer photography, the process of Polaroid film. Here the development process proceeds automatically after it has been initiated by pulling the film through rollers releasing the development chemicals inside the film.

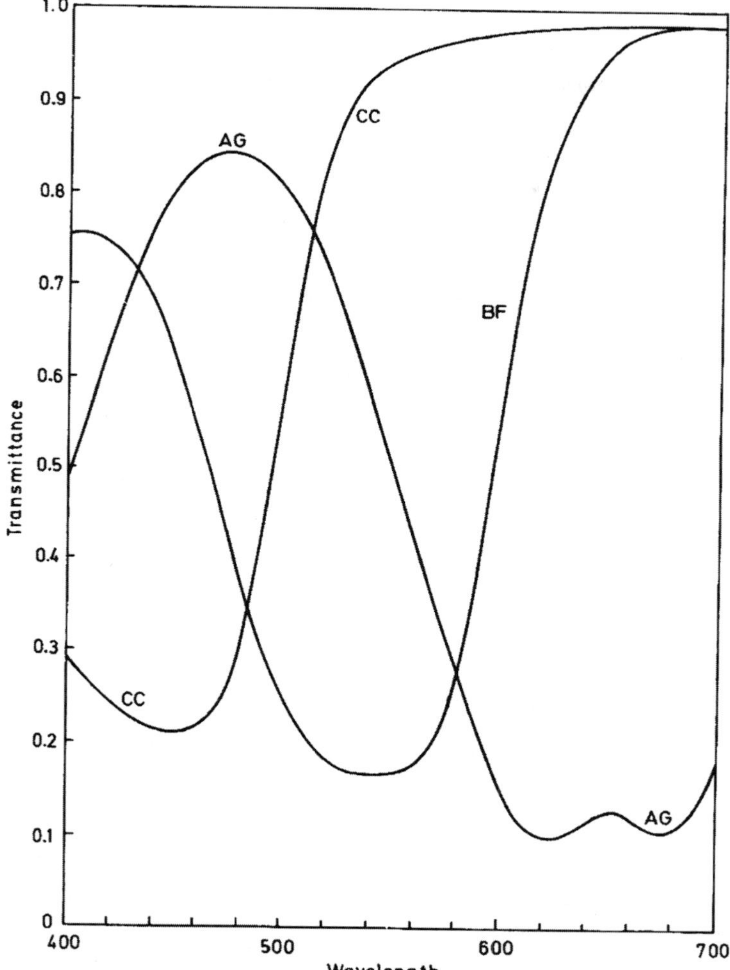

FIGURE 8.2 *Transmittance curves of the three dyes CC, BF, and AG of Figure 8.2 (4).*

PIGMENTS

As mentioned, pigments usually consist of crystalline particles of colorant molecules that have very low solubility in most solvents. Like dyes, pigments can be inorganic or organic in nature, natural or synthetic. Among the classical pigments are lead white (lead carbonate), minium (lead oxide), vermilion (mercury sulfide), ochre (various iron oxides), ultramarine, and indigo (note that the last appears both as a dye and a pigment). Pigments were either found as natural materials and processed or they were synthesized, such as minium, obtained by exposing lead white to high heat (5).

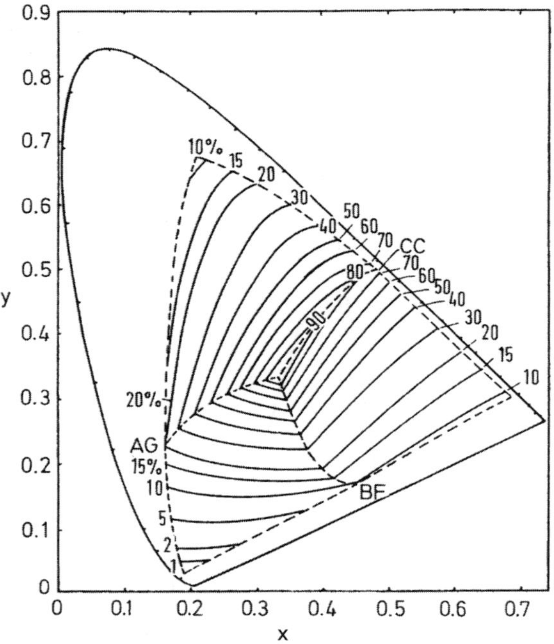

FIGURE 8.3 *Plot of chromaticities of three photographic dyes, yellow CC, magenta BF, and cyan AG alone and in two-dye mixtures of various ratios and concentrations on the dashed outline within the CIE chromaticity diagram. The solid contours represent planes of constant luminous transmittance; thus the three-dimensional shape of the complete gamut can be imagined (4).*

Ultramarine, pulverized and refined lapis lazuli from Afghanistan, required extensive processing to remove impurities, and the finest product was exceedingly expensive. Modern synthetic pigments are produced in multiple manufacturing steps and have comparatively high chemical purity.

To determine the interaction of light with pigments always requires consideration of the absorbance and scattering properties of the pigments themselves as well as, in case of translucent or transparent substrates, the corresponding properties. As with dyes, if they substrate is identical, it can be neglected in comparisons. Absorption is a function of the chemical constitution of the pigment. Scattering is a property that depends on the refractive indices of the scattering pigment and the medium surrounding it. This is illustrated with the case of ground glass powder. In air the powder is more or less opaque, indicating high scattering. Immersed in water it is translucent because the refractive indices of water and glass are similar.

The effect of the contributions of absorption, transmission, and scattering on the reflectance properties of pigments were described in 1931 in a comparatively simple two-beam model by the physicists Kubelka and Munk by considering the relationship in thin layers constituting, say, a paint layer (6). A schematic sketch of such a layer is illustrated in Figure 8.4. A fraction of the light arriving at the surface of the layer

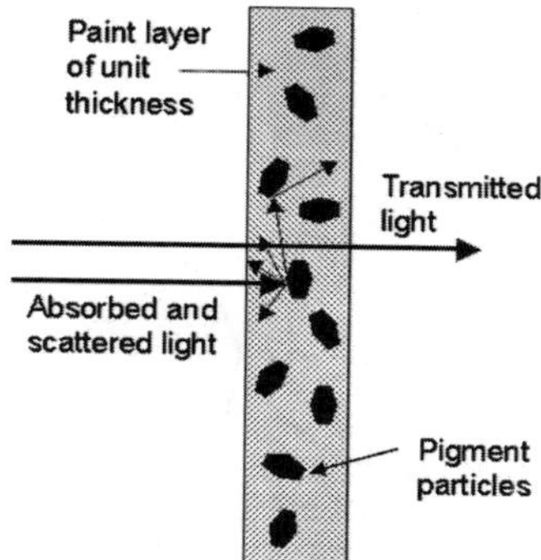

FIGURE 8.4 Schematic depiction of a unit paint layer indicating the incident light beam, absorbed, transmitted, and scattered light. Absorption is characterized by the spectral constant K *and scattering by* S.

is scattered internally by the pigment particles. Some photons are scattered forward, others sideways, resulting in multiple internal scattering, others are scattered backward and out of the layer. Some rays may not meet up with any pigment particles and are transmitted unscathed. Thus, there is a forward and a backward beam of light. Depending on the rays's wavelength and the absorption properties of the pigment particles, most rays are absorbed sooner or later in the layers. The Kubelka–Munk model defines spectral absorption and scattering constants for the pigments and a mathematical relationship between reflectance and the ratio of absorption constant K and scattering constant S. This relationship is considered additive, that is, in mixtures the spectral K/S values of the pigments involved, weighted by their concentrations, are added together with the K/S values of the substrate to predict the reflectance function of the mixture. Given the many variables involved and the complexity of the relationship, this simple model works surprisingly well in many situations. Much more complex scattering models have been developed but the determination of the corresponding multiple constants is difficult, and they need to be fine-tuned to specific narrow conditions (7). K and S spectral constants are expressed relative to scattering and absorption behavior of a white standard pigment (usually titanium dioxide) for which the relative values of K and S have been set at 1 across the spectrum. Figure 8.5 illustrates K and S values for an orange pigment. Its curves show that most light in the short-wave half of the spectrum is absorbed, while light in the other half is scattered, but only with half the efficiency by which titanium dioxide scatters light in the same region. To determine such functions requires the preparation of multiple samples of pigment mixtures with white and black standard pigments and the application of algebraic procedures (with the help of a computer) (8).

As mentioned earlier, in the case of dyes on textiles, scattering S is considered to be constant with a value of 1. The Kubelka–Munk ratio K/S thereby reduces to $K/1$ or K, the absorption of the dye. But since the underlying substrate scatters light, the Kubelka–Munk relationship is also used in this case rather than the

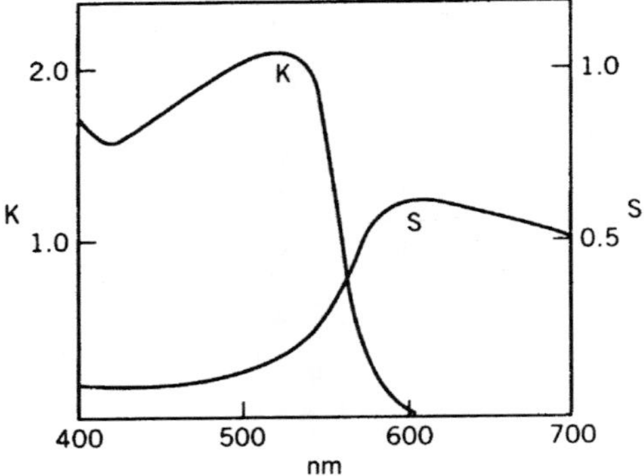

FIGURE 8.5 Spectral Kubelka–Munk K *(absorption) and* S *(scattering) functions of an orange pigment.*

Lambert–Beer relationship between dye absorbance and transmittance, used in the case of dye solutions. In practice, for both dyes and pigments there are often deviations from the Kubelka–Munk relationship having to do with variability in substrate and colorants, as well as physical interactions between colorants in mixtures.

During the manufacturing process, pigments are ground to optimize their particle size for absorption and scattering properties. Optimally ground particles result in pigments providing high coloration power as well as high opacity in a paint layer. Optimal particle size is a compromise between absorption and scattering. For optimal absorption, particles must be very small, ideally monomolecular, as in case of dyes. For optimal scattering, the particles must have a certain size that depends on the wavelength of light. The optimal particle size varies somewhat by pigment, and in practice pigments can only be ground to a range of particle sizes. Figure 8.6 schematically illustrates the relationship between particle size and scattering properties.

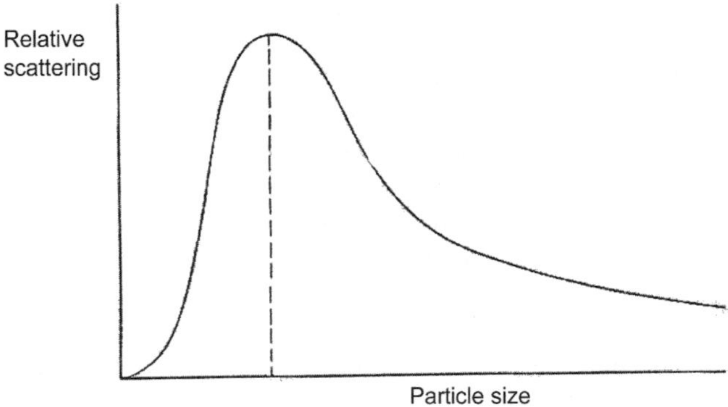

FIGURE 8.6 Schematic relationship between pigment particle size and scattering; the dashed line indicates the peak of scattering power.

COLORIMETRIC PROPERTIES OF COLORANTS

In Chapter 5 it was stated that the relationship between colorant concentrations and the resulting perceptions (of an average observer in controlled standard conditions) is not linear. If colorations of dyes or pigments at a range of concentrations are measured for reflectance and the resulting tristimulus values are plotted, for example, in the form of chromaticity coordinates, the resulting lines connecting data points are usually curved (Fig. 8.7). Such lines are called *colorant traces*. In the case of many dyes and particularly pigments traces usually do not end at the point of highest saturation, but at higher concentrations the colorations are seen as duller again (as well as darker). These are facts well known to painters. In pigments distinction is made between masstone, the concentration of the chromatic pigment in the medium where complete opacity is obtained, and results of the increasing replacement of the chromatic pigment with a white pigment (tint series) (10). In Figure 8.7 the resulting traces of 13 pigments,

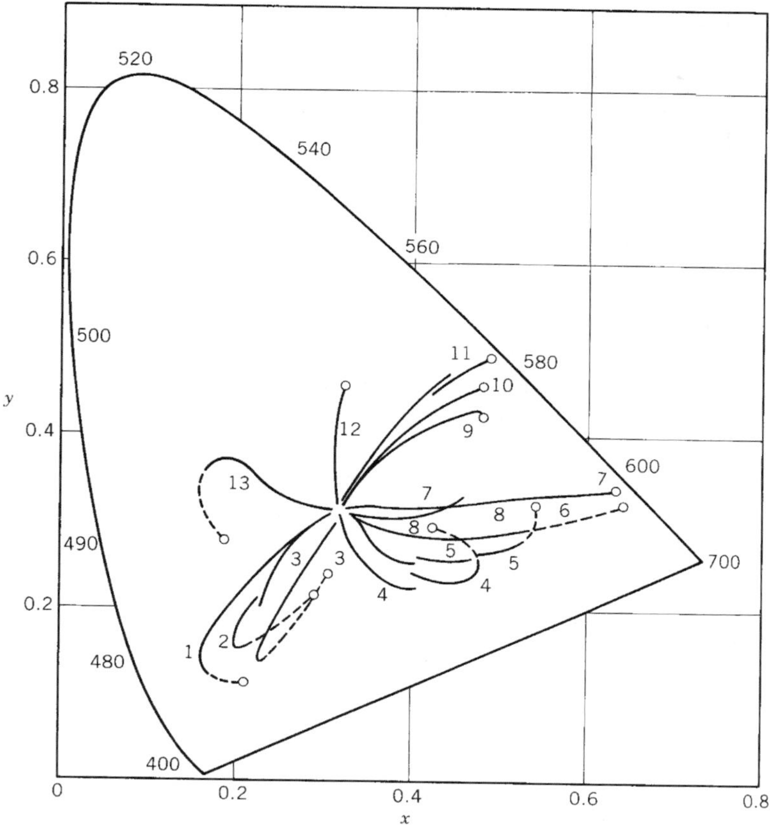

FIGURE 8.7 *Traces in the CIE chromaticity diagram of 13 commercial pigments (9).*

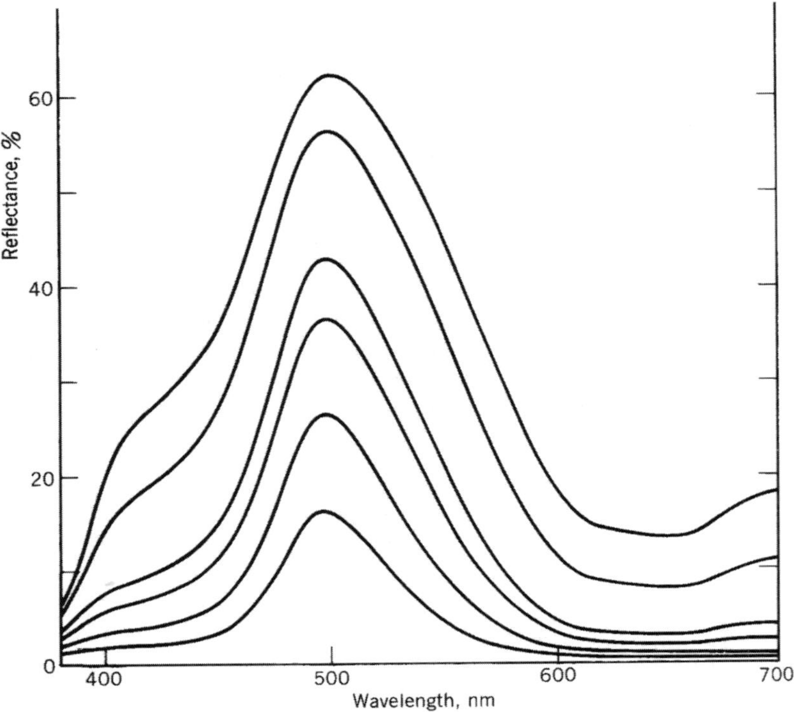

FIGURE 8.8 Reflectance functions of pigment #13 of Fig. 8.7 in masstone and various mixtures with titanium dioxide white pigment (11).

beginning at the masstone (open circle) and ending in the central neutral point, are shown. They indicate very varied behavior, including changes in chromaticness as well as hue. These changes are implicit in the changes in reflectance functions. Figure 8.8 illustrates reflectance curves of pigment #13 of Fig. 8.7 in various dilutions with titanium dioxide in a plastic resin. Of course, the colorations in traces also differ in luminous reflectance. Figure 8.9 is a schematic illustration of three traces in the x, y, Y stimulus space. The traces and their divisions have no simple relationship to perceptual attributes.

COLORANT MIXTURES

When mixing colorants the range of stimuli is expanded from lines in the chromaticity diagram to areas, or in the x, y, Y space to volumes (as shown in Fig. 8.3). When mixing two colorants in differing ratios the results vary in an irregular manner. When adding small amounts of a red or a blue dye to a yellow dye, the perceptual result is very large. On the other hand, adding small amounts of yellow dye to a red or a blue dye results in relatively small perceptual differences. The total range of stimuli obtained from the mixture of three particular colorants is, as mentioned, called its gamut.

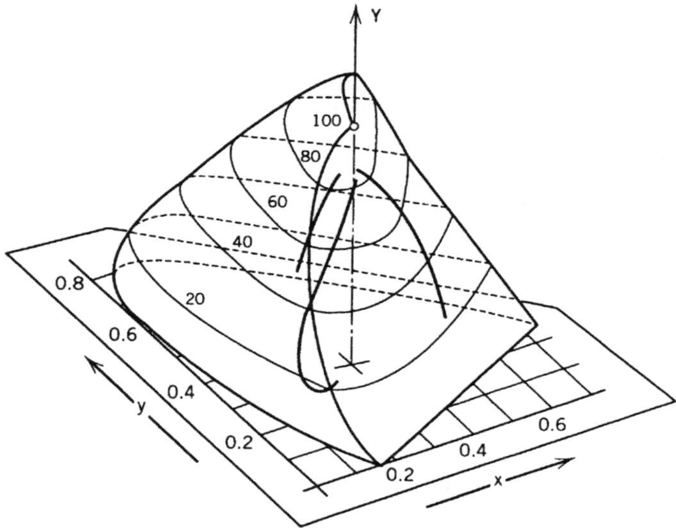

FIGURE 8.9 *Schematic representation of the traces of a yellow, red, and blue dye in the* x, y, Y *color stimulus space.*

It occupies a region of the optimal object color solid. These regions vary as a function of the optical properties of the three primary colorants. More on this subject is mentioned in Chapter 9. The chromatic gamut of three dyes is shown in the CIE chromaticity diagram in Figure 8.10. Total concentration on weight of goods of the Y, R, and B dyes in all mixtures is 2% on weight of material. The concentration ratios

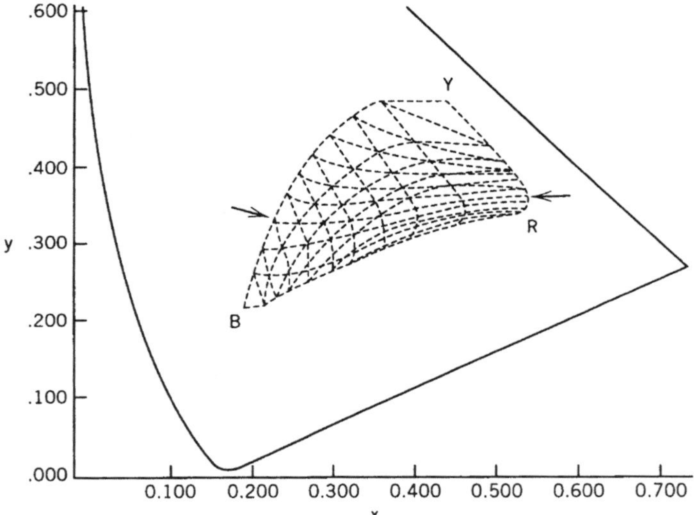

FIGURE 8.10 *Color loci of dyeings with mixtures of three dyes on a textile fabric. The total dye concentration in each dyeing is 2.0% on weight of fabric. Relative dye concentrations are varied in simple ratios (see text for more details).*

along the line indicated by arrows and starting on the left are as follows 7-0-3, 6-1-3, 5-2-3, 4-3-3, 3-4-3-2-5-3, 1-6-3, 0-7-3. To obtain a stimulus at a particular location in the triangle (in reality a three-dimensional space) the concentrations of the three dyes have to be appropriately adjusted. The figure illustrates the complicated relationship between dye concentration and stimulus. It also suggests the further possibility of determining colorant concentrations required for obtaining stimuli matching a standard.

SPECIAL COLORANTS

Fluorescent Colorants

Fluorescent dyes and pigments represent a special group among colorants. They consist of molecules with the property of light absorption in the near ultraviolet or visible region but, rather than reemitting a portion of the absorbed energy in the infrared region as is the normal case with colorants, they reemit energy at higher wavelengths in the visible range of the spectrum. Figure 8.11 illustrates the spectral reflection, emission, and sum functions of a 'red' fluorescent dye on a textile substrate. The amount of emitted light depends on the dye concentration, but also on the amount of light energy available for conversion and reemission. As a result, different light sources will result in different emission levels. At the wavelengths of emission the

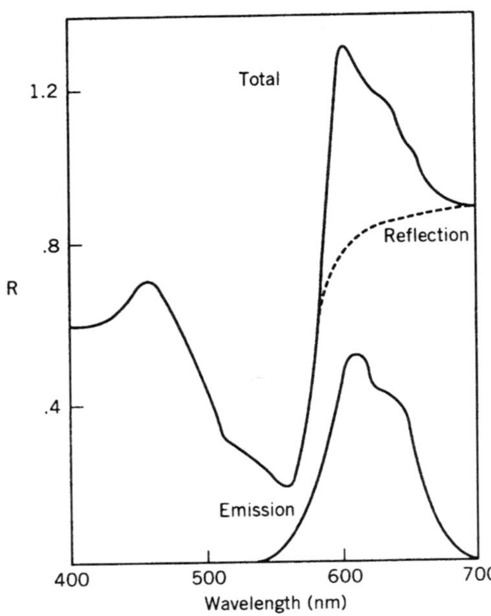

FIGURE 8.11 Reflectance, emission, and total sum curves of a fluorescent red dye on a textile material.

emitted light is added to the light regularly reflected there, resulting in light intensities exceeding that of the light falling on the sample. As a result, samples colored with fluorescent colorants tend to appear glowing, particularly in darker surrounds (think of a fluorescent yellow highlighter applied on white paper). They exceed the zero grayness level.

Use of fluorescent colorants either alone or in a mixture with nonfluorescent ones causes considerable technological difficulties in terms of stimulus definition. While today samples (particularly textile samples) are usually measured for reflectance with diffuse geometry where the sample is illuminated with broadband light separated into its spectral components only after reflection (see Chapter 6), fluorescent samples have to be illuminated with monochromatic or narrow-band light for correct separation of absorption and emission. In mixtures with other fluorescent or nonfluorescent colorants, there is a complex intermixture of absorption, reflection, and emission. Determination of tristimulus values (given appropriate reflectance measurement) follows the normal procedure. In the chromaticity diagram stimuli from fluorescent colorants must fall within the boundary of spectral lights. In the optimal object color solid, however, they can fall outside the boundary in terms of luminous reflectance because of emission of light. Some natural fluorescent colorants absorb energy in the near-ultraviolet region. Among them are a number of minerals and porphyrins, organic compounds causing, for example, the brownish color of brown eggs. Such eggs, illuminated by near-ultraviolet radiation, are perceived to have a glowing red color (12). Many natural products fluoresce, a fact made use of in crime scene investigation.

Fluorescent whitening agents form a special group of fluorescent colorants. They are used to make products such as textile materials, paper, and plastics appear whiter than they are. They absorb near-ultraviolet energy and emit visible light mainly in the range of 420–430 nm. They make white materials look whiter by masking with emitted light the yellowish tinge of some uncolored materials. The appearance of materials treated with fluorescent whitening agents depends strongly on the spectral power distribution at the beginning of the spectrum of the light in which they are viewed. In daylight, containing a fair amount of near-ultraviolet energy, they appear whiter than in the light of a tungsten lamp or many fluorescent lamps containing little or no ultraviolet energy. In pure near-ultraviolet light, they have a glowing bluish white appearance.

When fluorescent colorants are mixed with regular colorants, the total effect depends on the degree of overlap of absorption and emission bands. Emission can be completely absorbed by regular colorants absorbing in that region, depending on the relative concentrations.

Metallic, Pearlescent, and Interference Flakes

Inclusion of flakes of different metals and their combinations in paints has become a widely used practice in automotive coating. Flakes align themselves horizontally in the drying paint layer and cause internal reflection, giving the painted surface a lustrous metallic appearance. The exact effect depends on the metal used and the dimensions of the flakes. The appearance of such painted panels strongly depends on the angle of incident light and the angle of viewing. Paint technologists speak of the face and the flop color, the former (with incident light perpendicular to the sample) viewed at 45° to 60°, the latter at 75° to 110° angles. Many different specific effects can be obtained by manipulating components of the paint appropriately. Angular measurement of reflectance at several angles is required for quality-control purposes of such paints.

Pearlescent flakes are made from transparent materials with a high refractive index. When included in a paint medium, the result resembles the appearance of pearls or the interior of certain seashells. The most common modern material used as pearlescent (or nacreous) pigment is mica coated with titanium dioxide. The titanium dioxide deposits are in the form of platelets, resulting in interference colors (see Chapter 1).

A more recent interference color technology relies on deposition of thin films from metal vapors on inert materials. The final product has different layers: a metal layer on the surface of the inert material, followed by a glasslike layer and a semitransparent absorber layer. Such pigments exhibit dramatic shifts in color appearance as a function of the angle at which they are viewed. Different layer thicknesses and different absorbers result in a wide range of perceived colors. In combination with conventional pigments, such products can produce a wide variety of effects such as those seen today in nail polishes or automotive paints. Quality control of such materials becomes increasingly complex, requiring multiple measurements and knowledge of how ingredients affect the measuring results.

Colorants are convenient tools of technology for the modification of transmission and reflection characteristics of materials, thereby creating a wide variety of color experiences. This technology has a history stretching back at least 30,000 years. Today, there is a large variety of colorants available for coloration of materials, and new colorants appear on the market regularly. A wide gamut of surface color experiences can be achieved. Extensive application technology for dyes and pigments has developed over thousands of years. The material value of colorants is small compared to the commercial value they impart to the products to which they are applied. Colorimetry and the equipment necessary to apply it have aided substantially in the maturing of colorant application technology.

FIGURE 2.1 *A* Cornus kousa *tree with fruit. Left: black-and-white image; right: identical image in color.*

(a)

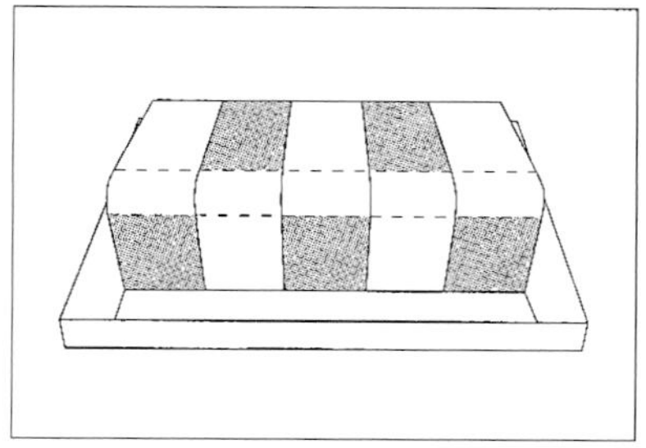

(b)

FIGURE 4.1 *(a) Computer-generated image of a box on a table top, illuminated from above and behind, before a varied background. (b) Fields colored in gray on the accompanying sketch have identical reflectance properties (metric lightness* $L^* = 54$*), but result in distinctly different appearance (3).* Figure also appears in color figure section.

FIGURE 4.2 *Examples of the Helmholtz–Kohlrausch effect. The three colored fields have the same metric lightness (CIELAB* L^**) as the gray surround, but appear noticeably lighter.*

FIGURE 4.5 *Example of simultaneous contrast. The five fields inserted into the varying surround are physically identical.*

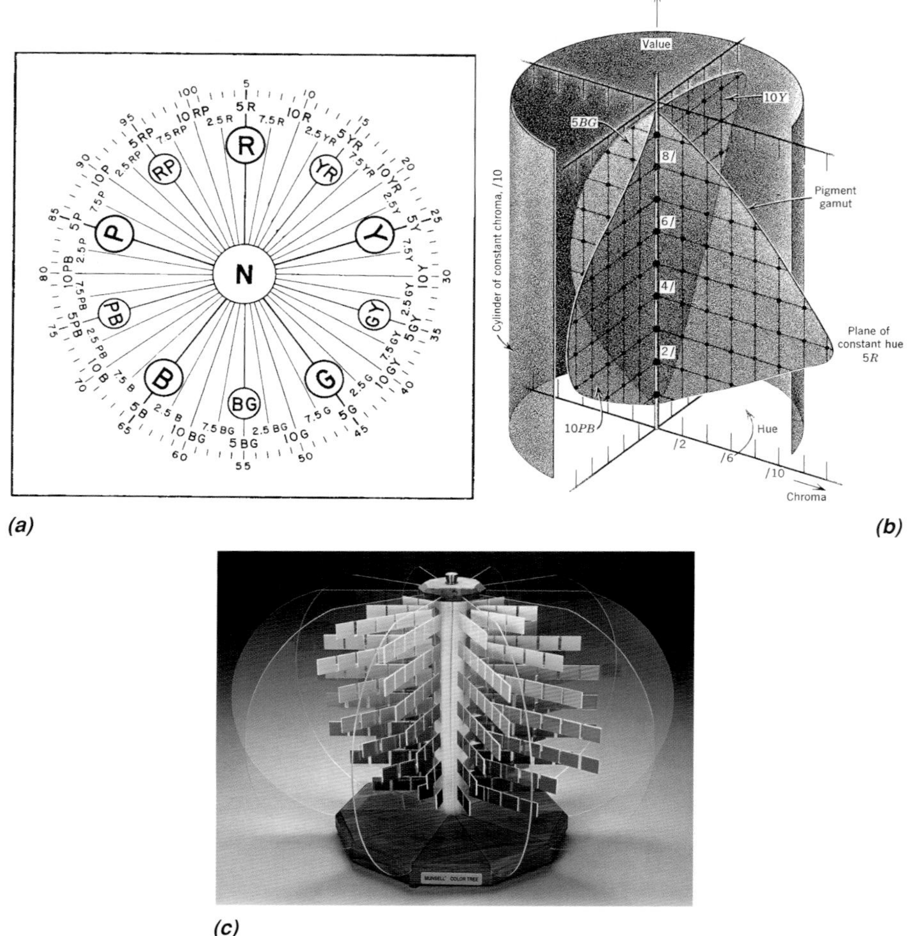

FIGURE 5.8 *(a) Organization of the Munsell color chart. (b) Conceptual illustration of the organization of the Munsell system (12). (c) Model of the Munsell Color Tree. (Image courtesy Gretag-Macbeth Corp.)*

FIGURE 5.11 *Model of samples of OSA-UCS, illustrating some of the cleavage planes of the system. (Image courtesy D. L. MacAdam).*

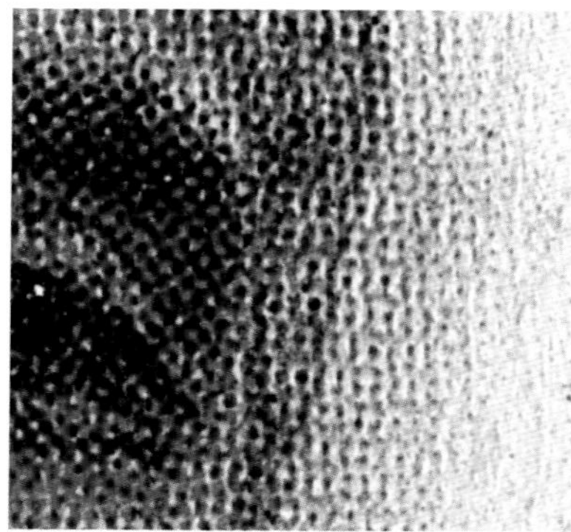

FIGURE 9.3 *Enlargement of a portion of a halftone reproduction of an art work showing the partially partitive and partially subtractive mixture of the primary pigments and black.*

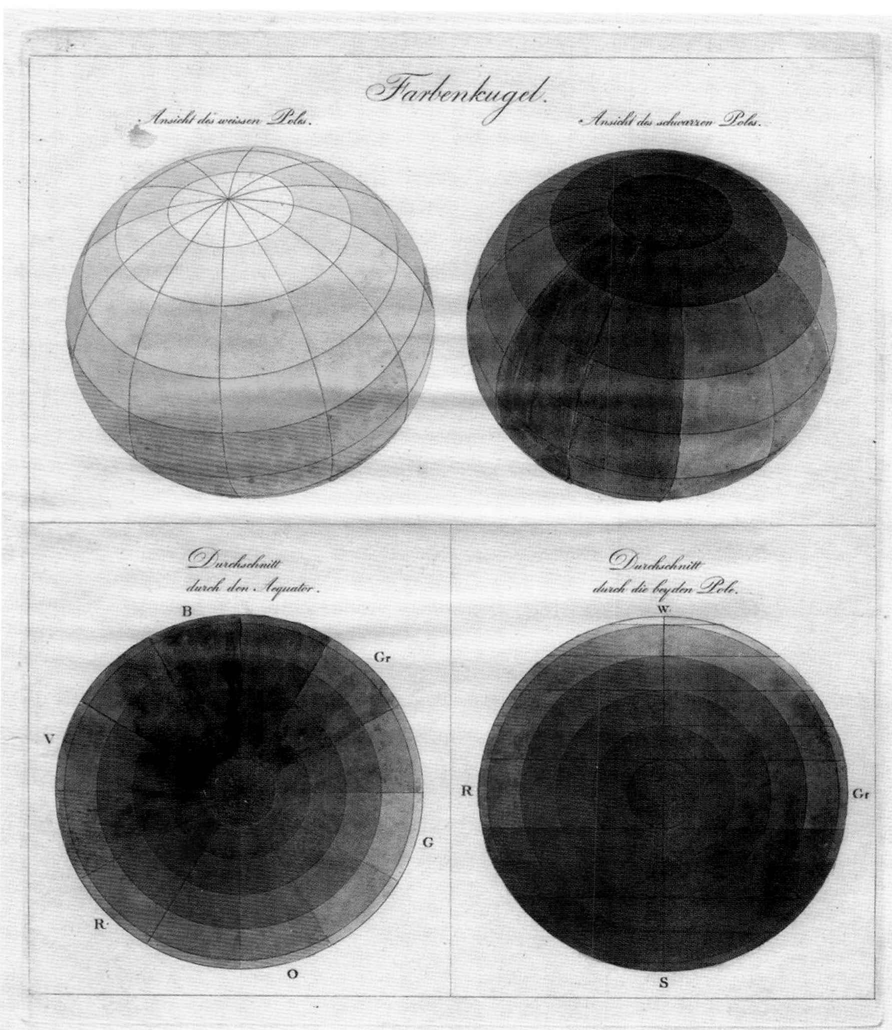

FIGURE 10.5 *Phillip Otto Runge's color sphere of 1810. On top are views toward the poles. On the bottom left is an equatorial section, and on the right a vertical section along the central axis. (Image courtesy Werner Spillmann Collection.)*

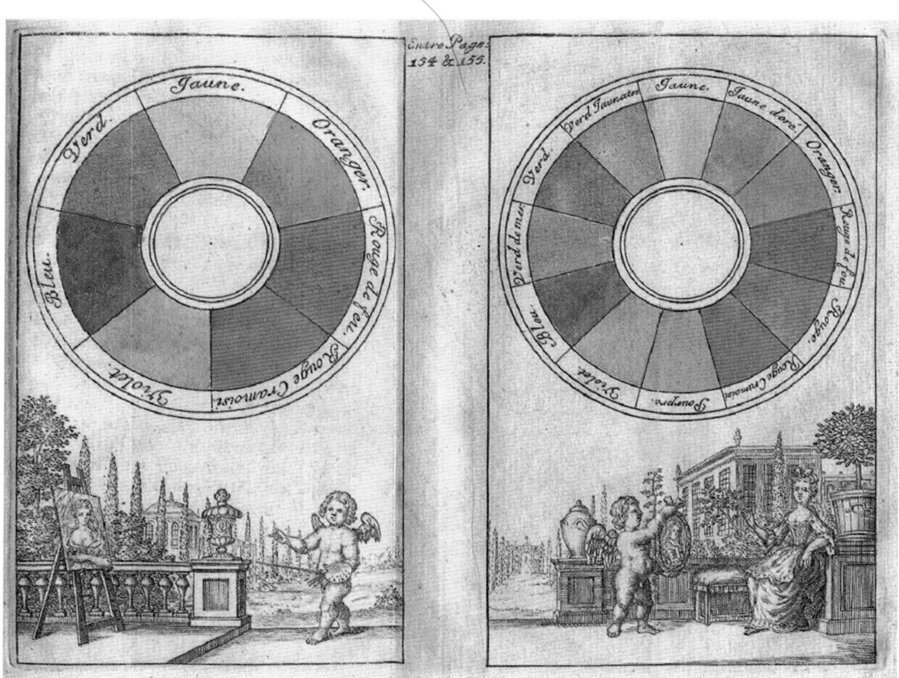

FIGURE 11.2 *The hand-painted color circles from the van Dole edition of 1708 of* Traité de la Peinture en Mignature. *(Image courtesy Werner Spillmann Collection.)*

9

Color Reproduction

Reproduction of color experiences from lights or materials in the same or another medium is an old problem. Some painters wish to recreate their experience of looking at a still life in as lifelike a way as possible. Some artisans want to duplicate for mass consumption the look of certain rare materials (say, the stone garnet) with inexpensive means (say, glass). Dyers may need to duplicate the color appearance of a fabric dyed by a competitor, or the painted standard of a customer. A photographer wants the child and flowers in the picture to appear as closely as possible to the original. A paint manufacturer may see sales opportunities with a line of paints duplicating colonial colors. A graphic designer wants to see the colors on the display screen of her computer reproduced exactly on paper. A large direct-mail retailer wants the customer to have an accurate impression of the colors of the merchandise shown in the catalog. The problems are manifold, the solutions often complex and specific to the problem at hand.

The general purpose of color reproduction is to have the appearances of original and reproduction match, preferably under all conditions. Since the media of original and reproduction are often different, an exact and nonmetameric match is generally difficult or impossible, as the spectral power distributions of lights or the reflectance properties of colorants tend to differ. In the case of exact reproduction of the spectral power of lights or spectral reflectance or transmittance and substrate of objects, identical appearance is guaranteed, but is rarely achievable.

A number of reproduction methods rely on a limited number of lights or colorants and on metameric matches of originals. Because of differences in the gamuts of lights or colorants used in various reproduction methods, the ranges of colors that can be

Color: *An Introduction to Practice and Principles, Second Edition,* by Rolf G. Kuehni
ISBN 0471-66006-X Copyright © 2005 John Wiley & Sons, Inc.

accurately reproduced varies in different systems. This is of consequence for stimuli resulting in bright, intense colors in the original medium that produce perceptions outside the gamut limit of the reproducing medium. In the second medium such stimuli will appear duller.

For some purposes exact reproduction is not desirable. In color photography there may be preferred coloration, either enhancing chroma in pictures of people or by use of polarizing lenses in nature photography, or lowering chroma and changing hues to achieve sepia tones. Energy-efficient fluorescent lamps (triband lamps) have been developed that for most objects result in higher perceived chroma than daylight, thus producing preferred coloration. Most consumer photography is today enhanced during the printing process to achieve pleasing results. It is evident that reproduction is possible at different levels of fidelity:

Spectral color reproduction refers to exact spectral duplication of the spectral signature of lights or objects; it results in highest fidelity. The match is valid for all observers and under all lights.

Colorimetric reproduction implies identical chromaticities and relative lightness or brightness. Because matches are usually metameric, high fidelity may only be obtained in certain lights and for certain observers; matches are conditional.

General appearance reproduction implies approximation of appearance when colored media differ, such as nature vs. color film, or monitor display vs. color printer. Here fidelity is usually limited.

Preferred reproduction refers to changes in appearance between original and reproduction to achieve pleasing results. Such changes can be obtained with the help of programs such as Adobe's *Photoshop* software, where hue, saturation, and brightness of the image can be manipulated, and blemishes and other extraneous object images removed or added. In this case, fidelity is not at issue.

Depending on the final application, any of these levels may be required or desired, but depending on the technologies involved, only one or two may be achievable. An example is the reproduction of artwork. Because of the technology involved (color photography, color printing, or computer display), matches of image components are virtually always metameric. Because we rarely have the opportunity to examine copy and original simultaneously fidelity, or lack thereof, of reproduction can generally not be judged. We only become aware of the problems when comparing reproductions in two different books or between different media.

Following are brief descriptions of four major technologies in which color reproduction is an important function and of the techniques used to achieve desired levels of reproduction.

Color Television

There are three different technologies that are widely used for displaying colored images. In conventional color television, the image is created by light emitted from three different phosphor compounds, resulting in red, green, and blue perceptions. Perceptions of other colors are created by partitive (additive due to size and proximity) binary or tertiary mixtures of points of light and by contrast effects. Electrical impulses from a controlled electron beam in a cathode-ray tube (CRT) excite dots of the phosphor compounds at different times and to different degrees, creating stimuli of varying spectral composition and intensity. The diameter of the electron beam determines the size of the dots. In computer monitors the number of dots (pixels) per unit distance is usually higher than on television monitors. In high-resolution monitors, there are as many as one hundred rows or columns of pixels per inch. A pixel represents a data point in the display and consists of three subpixels, red, green, and blue. The image is "refreshed," that is, the electron beam passes over the pixel areas about 60 times per second, resulting in uniform color appearance and the absence of jerkiness in moving objects.

Plasma display panels (PDP) are a special version of phosphor-emission-based displays. Here columns of phosphor compounds are sandwiched between two glass plates with spaces filled with a gas mixture, usually neon and xenon. Electric pulses can be sent via arrays of conductors to any location on the panel where they excite the gas plasma, resulting in local ultraviolet radiation exciting the nearby phosphor that gives off either red, green, or blue light in various amounts.

The third widely used technology is liquid crystal display (LCD). Here the light source consists of tiny fluorescent bulbs shining light through a one-way polarizer. Individual small areas in the display are turned on or off via transistors that cause liquid crystals to twist, letting white light pass through red, green, or blue filters, followed by another one-way polarizing filter. Color stimuli are generated by mixtures of various amounts of light that has passed through one or more of the filters. CRT and PDP have the advantage of brighter images than LCD. In PDP technology the image can be viewed without loss from a much wider angle than in LCD. As a result of the filter technique, the gamut of colors in LCD is noticeably smaller than in CRT or PDP.

The primaries used in color television are standardized, with standards differing in some countries. The U. S. standardizing body is the National Television Standards Committee (NTSC). The location of its primaries and the resulting chromatic gamut in the CIE chromaticity diagram are shown in Figure 9.1. The electric information creating the image on the display is either in analog or in digital form. In computer displays the information at the graphics-card level is digital. Intensity at the subpixel level usually can be controlled in 256 steps, resulting in 16.8 million different color signals (1).

Color Photography

As briefly described in Chapter 8, color film consists of layers, each containing a silver compound and a sensitizer tuned to one of three spectral regions. One kind of film, for transparencies, produces positive images, the other, for paper prints, produces negative images. After exposure the film is developed with dyes that attach themselves to the sensitized regions. The dyes differ for transparency and paper-print

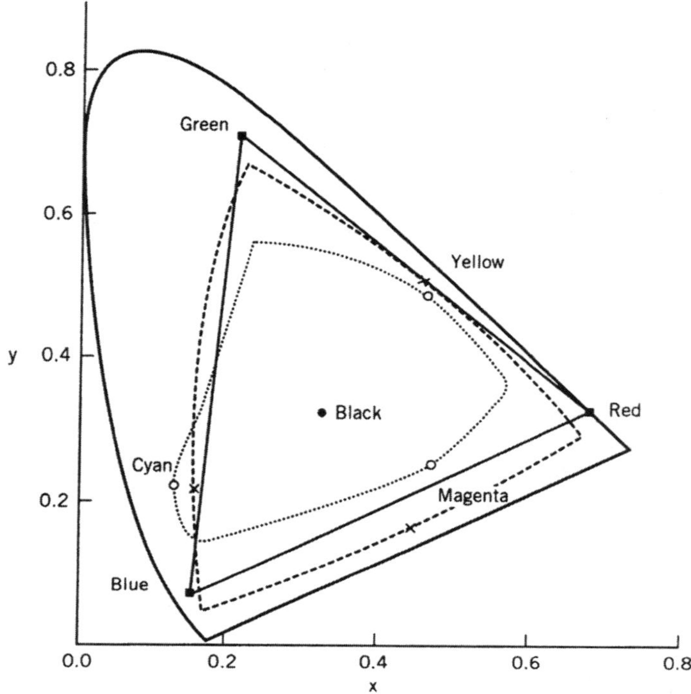

FIGURE 9.1 CIE chromaticity diagram with approximate locations and chromatic gamuts of color television primaries (——), color photography dye primaries (×-----×), and pigment primaries (plus black) used in halftone printing (○. . . . . . .○).

film. In the case of negative film, the image is printed in a second step where light is transmitted through the developed negative film onto sensitive photo paper. The paper development process generates localized attachment of yellow, magenta, and cyan dyes to the sensitized areas. When the finished images are viewed light is absorbed by the dyes in the layers and the remainder reflected from the white paper background, creating by subtractive mixture the final color stimuli. In the case of the transparency film, portions of the light beam passing through the transparency are absorbed and the remainder is reflected from a white screen, creating the image on the screen (2).

The transmittance functions of three typical photographic dyes were shown in Figure 8.2. They were selected for the narrowness of their absorption troughs (minimal overlap) and the fact that together they cover most of the visible spectral band, giving them a comparatively large three-dimensional gamut. Their location in the chromaticity diagram and the chromatic gamut are shown in Figure 9.1. Selection of dyes is tuned to the light source under which the subject matter is photographed. Specialized films are available for photographing in tungsten light to adjust for the chromaticity of that light, which the human observer adapts to automatically.

Color photography discussed so far is an analog process. The image can be digitized either immediately by using a digital camera or by scanning the analog image, and is then available for manipulation using computer technology and software. In this way, improved reproduction or preferred reproduction can be achieved. In digital cameras, the image is digitized directly on the image plane of the camera. The resolution depends on the number of pixels per unit area of the image plane. Pixels are again divided into red, green, and blue subpixels. Most digital cameras extract a measure of the quality of the light illuminating the subject and more or less well adjust the final image for it (3).

Graphic Printing

Most of the colored illustrations in magazines, newspapers, and advertisement flyers are printed in the four-color halftone process. The colorants used are pigments in the form of printing inks and, as in color photography, they represent subtractive primaries, yellow, magenta, and cyan, with the addition of a black pigment. They have broader absorption troughs, more overlap, and less reflection than the photographic dyes (Fig. 9.2). This results in lower chroma and lightness and a reduced gamut (see Fig. 9.1). A black pigment is used to improve the appearance of dark and black-appearing colors and to sharpen contours. In halftone printing, the color stimuli are created in part by a partitive mixture (primary pigments in small dots side by side) and by a subtractive mixture (overlapping dots of more or less transparent pigment layers) (Fig. 9.3). Variations in hue are achieved by varying the ratios of the pigments, in lightness and chroma by varying the size of the printed dots and overlap. Several

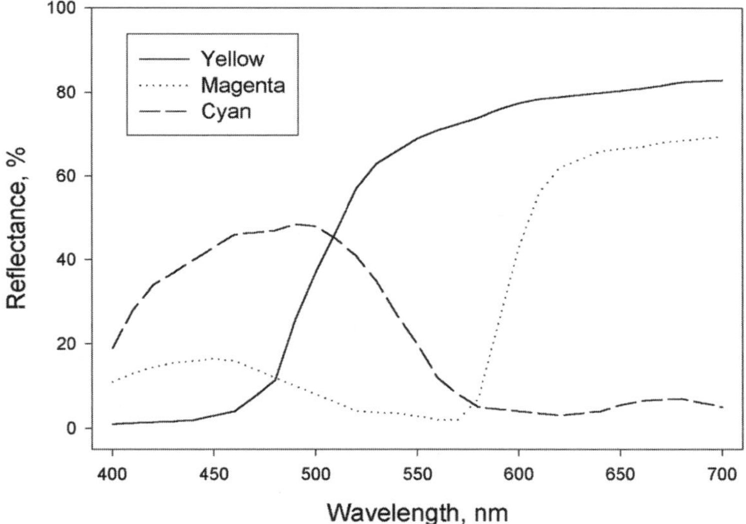

FIGURE 9.2 *Spectral reflectance functions of the three chromatic halftone printing primaries.*

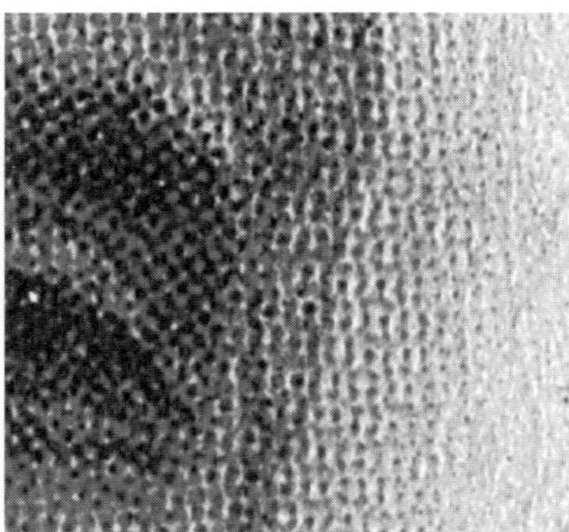

FIGURE 9.3 *Enlargement of a portion of a halftone reproduction of an art work showing the partially partitive and partially subtractive mixture of the primary pigments and black.* Figure also appears in color figure section.

proprietary techniques of halftone printing have been developed. The exact prediction of the results of such printing is complex because of the many variables involved (4). Using photographic techniques the color image is separated into four component images, each representing the quantitative information for application of one pigment. Also, in this case it was possible to improve color reproduction by digitizing the image and manipulating the data appropriately. For high-quality reproduction, the standard dot size is reduced and the gamut expanded by using up to seven different pigments.

For printing of extended uniform areas of the same color or a few well-delineated areas appropriately, mixed printing inks are used. Color reproduction in this case is aided by commercial systems, such as Pantone, that standardize printing inks to achieve a wide range of colors.

A specialized form of printing is that of digitized images generated on computer screens or in digital cameras. Several technologies have been developed, the most common today being ink-jet printing. Here microscopically fine droplets are projected in controlled streams from ink cartridges onto the substrate paper, where they dry and form the image. Depending on the quality of reproduction and the desired gamut, up to seven inks are used. Other techniques involve the xerographic process used in copiers or heat transfer and fusion of primary colorants onto the paper substrate.

Dyeing and Printing of Textiles and Paper, Coloring with Pigments and Paints, and Other Coloration Techniques

Textile materials or paper are generally dyed or printed to match a reference material. Wood is stained with dyes or painted with paints. Pigments are dispersed in plastic materials to color them. As discussed in the previous chapter, application of colorants changes the reflection (also scattering, in the case of pigments) properties of the substrate. In the case of paints, pigments are dispersed in a film-forming substrate applied to the surface of objects. Other specialized coloration techniques follow the same general pattern. In most cases, multiple colorants are required to obtain matching color perceptions. Depending on the requirements, a spectral match may be needed or a colorimetric match that is satisfactory for a limited number of light sources may be sufficient. If painting the walls of a room, initially a preference match to a reference sample may be desired. This changes to a spectral match for repair painting, particularly for critical situations, such as the repair of a car fender.

A large number of different colorants are commercially available for these purposes, as discussed in the previous chapter. These colorants vary in chemical and fastness properties light-absorption behavior and cost, and a great deal of expertise is required to find the appropriate products in a specific situation.

The highest degree of color fidelity is typically achieved in this field. It requires that identical colorants be used for reference and production, and in production at different locations, resulting in spectral matches. Frequently it is necessary, however, to find a compromise between resulting color fidelity, technical properties, and the economics of colorants. The problems become more complex when different materials, such as plastics, paints, or fabrics, need to be matched to have the same appearance. In the case of textiles, the problem is complicated by the fact that different fiber types require different chemical classes of dyes. Where the same dyes can be used on two different fibers, the resulting spectral absorptions tend to differ.

The requirements for color fidelity in these fields are often very high. The maker of garments, for example, wants to use pieces from different productions to cut panels for the same garments. In the case of repeat orders, leftover goods need to be used together with new goods.

COLOR MANAGEMENT

Today, there may be input of digital images from several different sources into a computer (keyboard, digital still or movie camera, scanner) and output onto several different devices (display units, projectors, printers). In order to obtain comparable images, "color management" is required.

A key aspect of color management is device-independent color encoding, that is, a method that normalizes the digital information independently of the particular properties of the device catching or displaying the information. Each pixel of the image is, in reality, represented by a complete spectral power distribution. For a complex image, this adds up to an enormous amount of information. Most digital devices reduce this information to three numbers per pixel, representing the implicit amounts of primaries. Different technologies and different manufacturers, for patent and economic reasons, may have somewhat different primaries. The situation is comparable on the output end: Most commercial computer printers have proprietary ink systems with different dyes or pigments, different drop size, and so forth, all adding variables to the results. In the computer itself, the information can be handled many different ways, depending on the capabilities of the computer and the programs run on it. The variety of results is infinite without some standardization. To stem this flood, the International Color Consortium (ICC) was formed in 1993 by the involved industry (with close to 100 member organizations and corporations). ICC has defined standards for the acquisition of images, reference medium, a color appearance model that translates the colorimetric data from one set of conditions to another, and colorimetric quality indicators, such as color difference calculation. Each proprietary input device can be characterized colorimetrically by imaging standard stimuli (such as the Kodak Color Checker chart) and transforming the results with the help of an appearance model into the so-called profile connection space (PCS), the reference colorimetric color space. Output devices also require characterization in terms of PCS. They can accept PCS data as input and convert it with their software to output in the proprietary space of the device.

A particular problem is the adjustment of different gamuts of devices. Figure 9.1 indicates that the chromatic gamuts of CRT and color printer differ significantly, and the question is how to scale data from one gamut into the other so that the result looks natural. This continues to be an area of significant research and commercial development. In different proposals, adjustments are made based on visual data and in others based on colorimetric transformations. Controlling the results of color reproduction in books, for example, requires a considerable process rarely possible in today's hurried and potentially worldwide production environment. But since direct comparisons between original and image are rarely possible, the reader is left to take the image as it is.

In most cases, variability in the processes requires preproofing of the results. Conventional proofing is expensive and time-consuming, so digital proofing methods have been developed. Both methods have somewhat limited accuracy, the former because of the variability between the proofing and the full-scale printing press, the latter because of the translations from one medium into another.

Terms like "reasonable" and "sensible" are used to describe the current results of color management (5). This indicates that they are less than perfect. It is unlikely that perfection is possible, given the nature of the human color vision apparatus and its variability (6).

COLORANT FORMULATION AND COLOR CONTROL

The classic method of colorant formulation is by trial and error and visual inspection of the result. This is the method that has been practiced over centuries and continues to be employed to a degree. Final approval of the formulation is still almost always based on visual inspection. What is required to achieve a reliable formulation is a uniform substrate, standardized colorants, tools to measure weights and volumes accurately, a laboratory coloration process that in its essentials duplicates the production process, standardized surround, and sources of light for the visual evaluation. It is also useful to have a collection of samples representing previous colorations from which the colorist can interpolate or extrapolate to the color that needs matching. The selection of colorants to be used depends on their "running properties" in the required application, on their fastness properties in the final product, and their economics. The colorist's personal experience with many colorants is useful. She determines a trial formula of usually three, sometimes less, and sometimes more, dyes and applies it in a laboratory process to a substrate sample. The result is evaluated under the required lights and, based on her experience, she makes adjustments to the formula. This process is repeated until a satisfactory result is obtained. The decision of when the approximation is sufficiently close is subjective, and two different observers may not agree on it. Knowledge of the final application of the colored material and the needs and desires of the customer may be important. In the process it may become necessary to replace one dye with another. Depending on the complexity of the requirements and the repeatability of the laboratory process, four to six or more trials may be required. Particularly difficult is the judgment of how to change the formula so that seemingly contradictory corrections in different lights can be achieved. This is obviously an expensive and time-consuming process, and efforts began in the 1940s to find technical support so as to arrive more quickly at an optimal formulation and acceptable result.

In many firms today, a technology known as *instrumental colorant formulation* is used to speed up the process. It requires a spectrophotometer, a computer, and the necessary software. Because of the problem of metamerism, most formulation software in use is based on matching the tristimulus values of the reference material (as viewed by a CIE standard observer under a CIE standard illuminant). The software requires access to colorant data in a form that is linear, or nearly so, against colorant concentration. The most commonly used linearizing function of reflectance is the Kubelka–Munk formula mentioned in the previous chapter. The colorants are taken to be additive, and the spectral Kubelka–Munk values (K/S) of the reference are taken to be the sum of the K/S values of the substrate and of the colorants used in the formula. For three colorants, this results in a set of three linear equations where the unknowns are the three colorant concentrations. The software must be able to assess if the calculated formula under the reference conditions leads to an acceptable result. A color difference formula, such as CMC, is used for this purpose. This formula is used to calculate color differences under various illuminants to assess the formulation's degree of metamerism against the reference. One of the advantages of instrumental colorant formulation software is that it can rapidly assess which colorants out of a group offer the optimal (for that group) formulation for matching the reference by attempting to calculate formulations for all possible combinations of three colorants from the group.

Figure 9.4 is a flow chart of a colorant formulation program. The reflectance function of the reference is measured or input in the form of numbers is provided, from which tristimulus values are calculated. Colorants are usually defined in the form of baskets containing products with compatible performance properties. The software then calculates the initial formulation by solving three simultaneous equations. If, as a result, the formula contains negative concentrations (as it would if an olive color is to be matched with two yellow colorants and a red one), it is dropped and the program advances to the next triplet of colorants. The program then calculates the reflectance function implicit in the formula from the unit K/S values of the colorants and their calculated concentrations and of the substrate, as well as the resulting tristimulus values. The color difference between reference and calculated formula is also calculated. If it is within set limits, the formula is accepted. If, as is more likely, the limits are exceeded, a different routine calculates in iterative steps successively improved

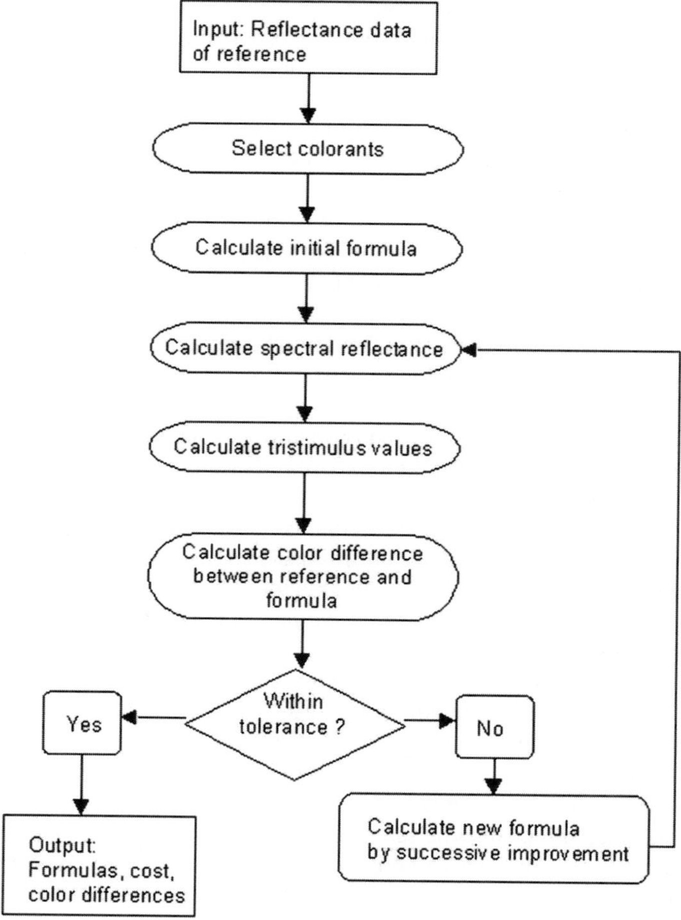

FIGURE 9.4 *Simplified flow chart of colorant formulation software.*

formulas until the quality criterion is met (or the formula abandoned after a set number of unsuccessful steps). For the final formula, color differences are also calculated under additional illuminants of interest as a rough measure of color shifts in these illuminants (a more exact method requires use of a color-appearance routine). After going through all possible combinations, the computer then displays all successful formulas with their colorant cost and color differences in the reference and additional illuminants. The colorist can now make additional judgments of the merits of the formulas and pick one or more for test colorations. It is unlikely that the first test produces an acceptable result because of a considerable number of process variables, colorant interactions, and the simplifying assumptions of the Kubelka–Munk theory. Similar calculations also can be used to determine formula corrections based on the first result. Such calculations determine the formula implicit in the coloration and compare it to the actual formula and make adjustments accordingly.

There is a considerable effort involved in making instrumental colorant formulation work well: the system represents a substantial investment in time and effort, and productive results are usually only obtained with considerable fine-tuning. With such efforts in place, usually substantial savings in time and colorant cost can be achieved compared to visual formulation. Formulation software programs continue to be improved and newer versions have self-learning features.

The same equipment can also be used for purposes of color control by making reflectance measurements and color difference calculations compared to the reference. Such calculations offer objective data with its advantages and limitations, as discussed in Chapter 7. In the case of textiles, there are two additional techniques used in color control. Textile fabrics are typically dyed in pieces, lengths of fabrics 50–100 yards in length. These may be dyed in the same machine or in different machines. As a result, each of the pieces representing a batch of a given color may have a slightly different average color, and the pieces need to be sorted so that no unacceptable differences between adjacent panels in a garment (for example) result. One technique to avoid this is shade sorting, where the pieces are sorted into "baskets" in which all pieces have acceptable differences. Another method is shade tapering, in which the pieces are sorted in a linear sequence based on magnitude and direction of difference, again so that pieces suitable for cutting into garment pieces are properly combined (7).

Depending on the substrate material, there are several other material characteristics aside from color that affect the final appearance, such as gloss, opacity, haze, surface structure, and content of metallic or pearlescent particles, that make appearance a function of angle of view and position of light source. This applies not only to observers but also to measuring instruments. Established relationships are usually only applicable to relatively narrow conditions that must be observed if a reasonably close relationship between observations and measurements is expected (8).

Color reproduction is a complex technological field of considerable importance. The problems that need to be solved are usually rather specific in regard to the colorants and coloration methods used. In practice, color fidelity ranges from poor to near-perfect. Generally, for a variety of reasons, perfect reproduction is impossible. In many cases, color is used solely for the purpose of obtaining attention or increasing discrimination, and exact reproduction is immaterial. It is a general rule that process costs increase with increasing color fidelity. Keeping fidelity moderate may be an economic necessity. At times color fidelity in reproduction is sacrificed in the interest of preferred coloration. The psychology of color vision, colorimetry, and color science in general have aided substantially in the refinement of production methods and will continue to do so in the future. But it is obvious that in many respects color reproduction technology is and will remain an art.

10

The Web of Color

Human ideas concerning the nature, purposes, and uses of color form a densely interwoven historical web. In the current chapter, a brief history of these ideas is presented with the primary focus on general and scientific aspects. Some philosophical ideas were presented in Chapter 2. Chapter 11 is a brief historical review of the development of color and color theory in art, with the special subject of color harmony presented in Chapter 12. The chronological development of these ideas is illustrated in the timetable following Chapter 12.

When humans became aware of colors is unknown. It must be assumed that it was before they began to give names to them, that is, well before about 7000 years ago, the beginning of writing. The genetic development of rods and cones indicates that early modern humans had essentially the same unconscious and conscious response to color stimuli that we have today. As for most people still today, color was simply recognized as an aspect of a material, helping to distinguish it from other materials. It was a valuable, if unrecognized tool in the daily struggle for survival. The application of names to color perceptions indicates a first degree of abstraction of color as a category, an entity separate from materials. Different materials can have the same or different color.

Experiments have shown colors to be perceived more quickly in the right hemisphere of our brain, but verbalized nearly exclusively in the left hemisphere by the average right-handed person. This indicates a fundamental difference between perception and verbalization, but one that we are not usually aware of.

The first presumably conscious use of colorants in the form of colored earths and wood cinders dates back some 30,000 years to a time when inhabitants of southern

Color: *An Introduction to Practice and Principles, Second Edition*, by Rolf G. Kuehni
ISBN 0471-66006-X Copyright © 2005 John Wiley & Sons, Inc.

Europe began to paint colored images on cave walls and to use colorants in early burial rites. Most of these images are naturalistic depictions of animals. In the earliest known site, the Chauvet cave, where work has been dated from 31,000 to 24,000 B.C., there are animal images engraved into the walls, others drawn in bold outline with charcoal, some with charcoal shading, others in red ochre with shading. There are positive handprints, collections of red dots, and even the image of an owl (1). In other caves, such as Altamira in Spain, there are bold outlines of animals in dark color filled in with lighter colors. Such paintings are believed to have had magical purposes.

Later, drawings were made outside of caves on rock walls, with often more abstract content, lines and symbols with unknown meaning and purpose, but also depictions of human activities. Such drawings are found from Africa to northern Europe, advancing northward behind the melting glaciers of the Ice Age, as well as in the Middle and Far East, and in the Americas. Egyptian artifacts indicate the use of colored pigments in decoration of clay figures at least 6000 years ago (2).

In Western culture, a curiosity about the nature of colors is documented first (as are many other things) in the writings of Greek philosophers. Colors, as the result of a perceptual process, may be the most important sensory information for humans (it is different, for example, for dogs, who have modest vision capabilities but a sense of smell greatly exceeding that of humans). Given this importance, it not surprising that philosophical consideration of the nature of color began very early.

GREEK IDEAS ON COLOR

The sixth century B.C. philosopher and mathematician Pythagoras believed Earth to be part of a perfect cosmos with a basis in numbers. This basis is symbolized by the tetractys, the triangle formed by representing the first four numbers as dots, adding up to 10 (see Fig. 12.1). The same numbers are involved in octave, fourth and fifth of musical harmony, believed to have been discovered by Pythagoras. According to Philolaus, Pythagoras equated colors with the next number, five. Plutarch quoted the views on color of the followers of Pythagoras as follows: "[They] called the surface of a solid *chroma*, that means color. Also, they named the species of color white, black, red and yellow. They looked for the cause of the differences in color in various mixtures of the elements, the manifold colors of animals, however, in their nutrients as well as the climatic regions" (3). Pythagoras is reported to have believed vision to be the result of a hot emission from the eyes, the so-called emission theory.

Another theory of vision was proposed by the fifth century B.C. philosopher Empedocles. According to him, energy flowing outward from the eyes and energy flowing from objects to the eyes intermingle and cause the sensations of light and color, the emission–immission theory. The classical elements fire and water are located in the eye, while wind and earth are located outside.

Democritus, flourishing in the second half of the fifth century B.C., was, like his mentor Leucippus, an atomist and the reported author of the comment: "By convention there is color, by convention sweetness and bitterness, but in reality only atoms and void." His four elementary colors were white, black, red, and yellowish green

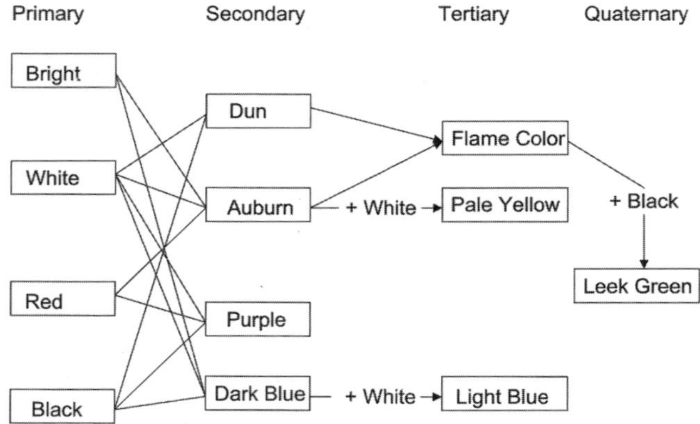

FIGURE 10.1 *Schematic representation of Plato's color-mixture scheme.*

(*khloron*). The other colors he said were derived from mixtures of the four primaries. He believed colors to be effluences from objects flowing to the eyes, according to the so-called immission theory (4).

Plato was critical of Democritus and a supporter of the emission–immission theory of vision. He produced an elaborate scheme of how colors are generated from an appropriate mixture of the four primary experiences bright, white, red, and black (see Fig. 10.1) (5).

Plato's student Aristotle, in the fourth century B.C., described in *Sense and Sensibilia* a scale of seven simple colors that became influential for the next 1700 years (see below). He compared flavors and colors and claimed each has seven simples. For colors he named black (*melanon*), yellow (*xanthon*), white (*leukhon*), crimson (*phoinikoun*), violet (*alourgon*), leek-green (*prasinon*), and deep blue (*kuanoun*), and believed that all others are derived from these by mixture. He had sympathies for the Pythagorean ideas and believed that particularly pleasant colors were the result of mixtures of primaries in the same ratios described by Pythagoras for the combination of musical tones. In regard to the nature of color, Aristotle developed a complex system of causative relationships to the four elements and four humors. Aristotle's disciple Theophrastus is believed by some to have been the author of the essay *On Colours* generally ascribed to Aristotle. Here we find the famous statement claiming that all colors derive from the simple colors black and white. Here are also statements that can be read as describing the color attributes of hue, saturation, and brightness.

In regard to color, Roman writers did not add anything of substance to the ideas of the Greek philosophers. The poet Lucretius wrote a verse version of Epicurean philosophy (*De rerum natura*, The way things are), in which he essentially followed the atomic theory of Leucippus and Democritus, including the immission theory of vision, as it was stated by Epicurus, but without explicitly mentioning colors.

Classical Greece invented many color words. Those used by epic writers add up to some 140, while the philosophers limited themselves to approximately 50 (6). The exact meaning of these words has been disputed and continues to be disputed. This

TABLE 10.1 Aristotle's Seven-Color Scale and Translations/Interpretations

Aristotle	Bartholomew	Aquinas	Trevisa	Dolce
Leukon	*albus*	*albus*	white	*bianco*
Xanthon	*glaucus*	*flavus*	yolow	*violato*
Phoinikoun	*puniceus* (*citrinus*)	*puniceus id est rubeus*	citryne	*croceo* (*giallo*)
Alourgon	*rubeus*	*alurgon, id est citrinus*	rede	*vermiglio*
Prasinon	*purpureus*	*viride*	purpure	*purpureo*
Kuanoun	*viride*	*ciarius, id est color caelestis*	grene	*verde*
Phaion/melan	*niger*	*lividus/niger*	blak	*nero*

dispute begins with Aristotle's list of seven basic colors. Translations of the extant works of Greek philosophers into Arabic, Hebrew, and eventually Latin did not clarify the matter. The list of the seven Aristotelian simple colors in Table 10.1 is used to provide an impression of the results. Some of Aristotle's works were translated into Latin in the twelfth century by William of Moerbeke, translations used by, among others, Thomas Aquinas. Quotations of Aristotle are found in many works, including the encyclopedic work of the twelfth century Franciscan Bartholomew the Englishman *De proprietatibus rerum* (On the properties of things). This very popular book was translated into Old English by John Trevisa, circa 1490, and into more modern English by Stephen Batman in 1582 (7). There are several quotations of and commentaries on Aristotle from the Italian Renaissance, for example, those by Lodovico Dolce in his dialog on color (1565) (8). Table 10.1 lists the presumed original words by Aristotle and the result of various translations/interpretations.

There are several remarkable facts in this table. It is generally assumed that Aristotle's list is strictly in lightness order, as it begins with white and ends with black. Aristotle did not specifically claim that, and his understanding of the color terms he used is not known. *Xanthon* is generally regarded as meaning yellow, *phoinikoun* (Phoenician) has been interpreted to refer to the reddish brown color of fruit of the *phoenix* date palm. *Alourgon* is usually understood to have the meaning of bluish red to violet, and *prasinon* that of leek-green. *Kuanoun* is generally regarded as having the meaning of dark blue, *phaion* as gray, and *melan* as black. Bartholomew appears to have initiated the use of the latinization of the Greek word *glaukon* to have the meaning of yellow, also used by the Franciscan Roger Bacon in the thirteenth century (9). More often, the meaning of *glaukon* is taken as bluish gray. Bartholomew translated *phoinikoun* into *puniceus*, but equated it with orange, *alourgon* became red, located in the center of the scale. *Prasinon* and *kuanoun* were obviously switched. He edited Aristotle's scale to fit his purpose, to have red in the center and two colors each between white and red and black and red, elevating red to the key position among chromatic colors. He may have switched green and purple because the latter fits better into the sequence yellow, orange, red, purple. In the late Middle Ages, it became an accepted fact that red was in the center of Aristotle's scale, halfway between white and black, perhaps because of the wide availability of Bartholomew's book in various editions and translations into several languages. Aquinas stuck closer to the original, in this respect, but he equated *puniceus* with *rubeus* (red) and, surprisingly, *alurgon* with orange. He latinized *kuanoun* as *ciarius* and called it sky blue. The Trevisa translation follows Bartholomew closely. Dolce was apparently influenced by Bartholomew, but for obscure reasons translated the first chromatic color as *violato* (violettish), and red has become the yellowish red of vermilion. In the absence of standard materials, the meaning of color words was (and is) largely up to interpretation.

Aristotle's (or perhaps Theophrastus's) idea that colors are generated from mixtures of white and black proved influential until it was replaced by Newton's experimental results on the splitting of daylight by refraction. But some people, including the poet and natural philosopher Goethe, remained unconvinced of Newton's results. In a general, metaphorical way Aristotle can be defended. Color stimuli can be seen as spectrally selectively "shaded" versions of daylight.

MEDIEVAL AND RENAISSANCE THOUGHT ON COLOR

The idea of linear color scales based on brightness was revived and quantified to some degree in the Renaissance. In the fifteenth century, the Italian philosopher Marsilio Ficino proposed a linear scale starting at black, followed by seven chromatic colors, with white at level 9, transparent or shining at level 10, brilliant at 11, and concluded the scale at level 12 with splendor. A more quantitative seven-grade scale of colors was described in 1563 by the Italian physician Hieronimo Cardano: "... we assume that white contains a hundred parts of light, scarlet fifty, black nothing." Yellow is described as having 65–78 parts of light, green 62, deep green 40, wine color 30, blue 25, and blackish gray 20 (10).

Painters, dyers, and producers of colorants had known long ago that there is much more to colors than a seven-grade Aristotelian scale. Three tint/shade scales, ranging from white to the undiluted pigment, and from there via mixtures with black to solid black, were described in the eleventh century by the Persian philosopher Avicenna. In the twelfth century, the Spanish–Arabic lawyer and philosopher Averroës introduced terms later translated into Latin as *remittere* and *intendere*, with the meanings to yield, abate, respectively, to spread, move toward. These have been interpreted as expressing the idea of such tint/shade scales. Explicit tint/shade scales from white, via the full pigment color, to black were first described by the German Benedictine monk Theophilus in his treatise *De diversis artibus* (The various arts) of approximately 1122, a compilation of methods and recipes for painting, glass making, and metalworking (11). He described such a tint/shade scale of the pigment vermilion as follows: "Then mix from vermilion and white whatever tones you please so that the first contains a little vermilion, the second more, the third still more, the fourth yet more, until you reach pure vermilion. Then mix with this a little burnt ochre, then burnt ochre mixed with black and finally black ... You can never have more than twelve of these strokes in each color range. And if you want these many so arrange your combinations that you place a plain color in the seventh row."

Basic understanding of the phenomena of color had not advanced by the end of the sixteenth century, perhaps because there were so many different kinds that

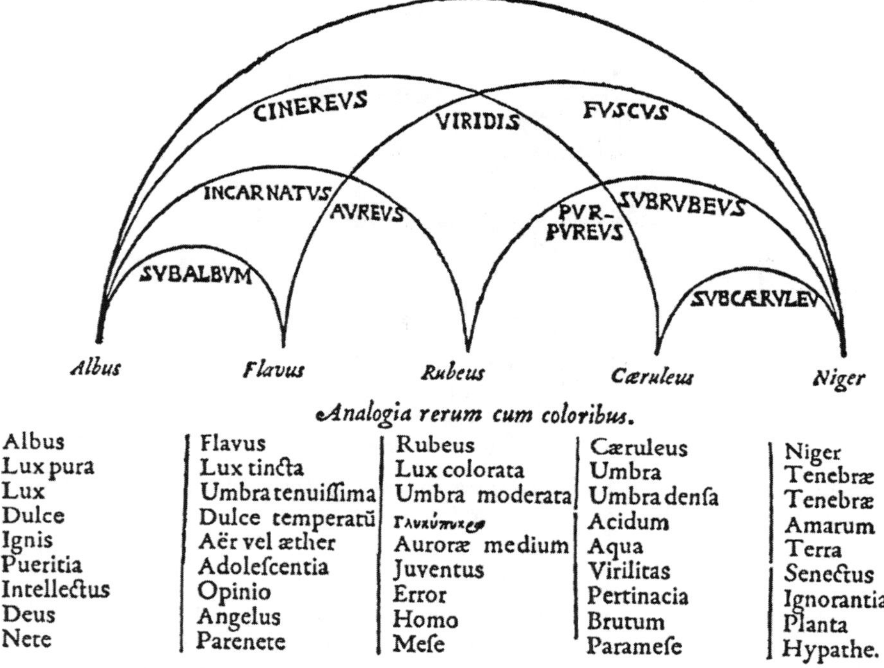

Albus	Flavus	Rubeus	Cæruleus	Niger
Lux pura	Lux tincta	Lux colorata	Umbra	Tenebræ
Lux	Umbra tenuissima	Umbra moderata	Umbra densa	Tenebræ
Dulce	Dulce temperatũ	γλυκύπικρος	Acidum	Amarum
Ignis	Aër vel æther	Auroræ medium	Aqua	Terra
Pueritia	Adolescentia	Juventus	Virilitas	Senectus
Intellectus	Opinio	Error	Pertinacia	Ignorantia
Deus	Angelus	Homo	Brutum	Planta
Nete	Parenete	Mese	Paramese	Hypathe.

FIGURE 10.2 *Athanasius Kircher's color scheme of 1671.*

needed to be explained in an all-encompassing manner, impossible before some clear facts had been established. Single phenomena, such as the rainbow, had found optical explanations that proved valid, in this case, apparently first by the German Dominican friar Dietrich von Freiberg in approximately 1310. Writers on the general subject of color continued to base their accounts on the classical sources. Among these were the Jesuit priests François d'Aguilon and Athanasius Kircher in the early to mid-seventeenth century. Both used arc diagrams of a type invented by the late Roman philosopher Boethius in the fifth century for representing facts of logical and musical tone relationships (Fig. 10.2). The idea of the general harmony of the natural world made it obvious to also use a diagram employed to express musical harmony for colors. D'Aguilon limited the basic chromatic colors to three: yellow, red, and blue. In his figure Kircher connects these in pairs as well as with black and white, thereby showing hue and tint/shade scales, as well as a gray scale. Medium mixed colors are identified in the loops. In the spirit of his time, Kircher placed eight series of analogs in a table under the five basic colors, ranging from light, via taste, ages of man, a scale from God to plants, to the strings of the Greek lyre. Kircher quoted in his discussion of the essence of color primarily the Greek authorities. In another chapter, he discussed the colors of angels (12).

A diagram of a gray and four linear tint/shade scales was drawn by the Finnish astronomer Sigfrid Aronus Forsius in a 1611 manuscript on physics (Fig. 10.3).

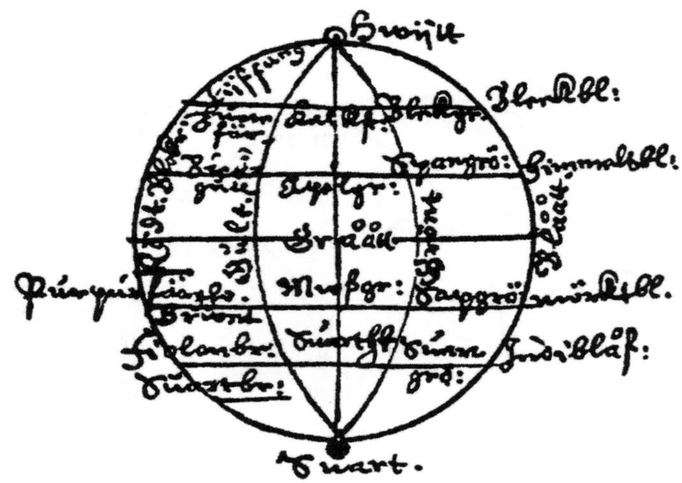

FIGURE 10.3 *A gray (vertical line) and four tint/shade scales connected to common white and black points by Sigfrid Aron Forsius, from 1611.*

He connected these to common white and black: on the left are red and yellow (gold) scales, in the center a gray scale, and on the right green and blue scales. Unfortunately, his manuscript was never published and had little influence on further developments (13).

An interesting contribution to the subject of color order was made in 1677 by the English physician Francis Glisson, who inserted in his last medical book a chapter on hair color and described there a color-specification system. This system made it possible to express any color as a combination of grades of a gray and three chromatic color scales: yellow, red, and blue. The scales were constructed in a novel manner to be perceptually equidistant. For the gray scale Glisson determined the midpoint between white and black as consisting of a mixture of 600 grains of lead white and 12 grains of carbon black. For the next lighter grade, he used 650 grains of white and 11 grains of black, and continued in this manner in both directions toward white and black. The three chromatic scales were constructed in a similar fashion, but only from white to the undiluted full color. Glisson believed that he could judge the grayness/blackness of darker colors using the gray scale. The resulting scales are approximately perceptually uniform, as a reconstruction has shown. Glisson rated the color of a golden yellow blossom as grade 11 of the yellowness scale, grade 3 of the redness scale, and grade 2 of the gray scale. In similar fashion all other colors can be expressed as the sum of grade values (14). A different step toward standardization was taken in 1686 by a member of the Royal Society of London, Robert Waller. He prepared a table of 1:1 mixtures of well-known pigments with white, yellow, and red pigments on the vertical axis, and blue and black pigments on the horizontal axis. Most resulting mixtures were identified by names in three or more languages.

The English physicist and chemist, Robert Boyle, succinctly expressed the general dissatisfaction with the state of knowledge on color in his 1664 book *Experiments and Considerations Touching Colour*. He classified the major schools of thought existing at that time as follows:

1. Aristotelian school.
2. Platonic school.
3. Atomistic school.
4. Kircher's school explaining color from mixture of light and shadow.
5. Chemical school explaining colors as caused by the alchemical three elements sulfur, salt, and mercury.
6. School of Descartes, explaining the sensation of light by the impact of light particles on the optic nerve and colors by the differing speed and movement of the light particles.

Boyle himself believed black and white to be the result of different degrees of reflection but did not consider himself competent to explain chromatic colors.

THE REVOLUTION OF THE PRISM

By the mid-seventeenth century crude prisms made from glass or quartz crystals were well known and had been used as toys and experimental tools. Among those known to have experimented with prisms are the thirteenth century Polish natural philosopher Witelo, Descartes, Boyle, and the Italian Jesuit priest Francesco Maria Grimaldi, who wrote a text, *Physico-mathesis de lumine, coloribus et iride* (Physical classification of light, colors, and the rainbow), published posthumously in 1665. This book represents an entirely new type of scientific experimental description and logical deductions from the results. Among his experimental results, he described the spectral composition of light (15).

At the time of publication of Grimaldi's book, Isaac Newton was 23 years old. Newton conducted his first experiments with prisms in 1666 and had a presentation of his findings read to the Royal Society in 1671, wherein he summarized them in thirteen propositions. These findings were published a year later. According to his propositions differently refrangible rays of light exhibit different colors. Colors and refrangibility are always connected in the same way. The color of a given ray cannot be changed by further refraction or reflection. When mixed, they "constitue a middling colour.... There are therefore two sorts of colours, the original and simple ones and others compounded from these." As simple colors, he mentioned red, yellow, green, blue, and violet-purple, "together with Orange, Indico and an indefinite variety of Intermediate gradations." These colors can also be obtained by composition. White is the result of composition and is not represented by a single ray. He concluded "that the Colours of all natural Bodies have no other origin than this, that they are variously qualified to reflect one sort of light in greater plenty then another" (16). In one masterful stroke the 24-year-old Newton clarified important aspects of the composition of sunlight that had been a mystery up to then. Newton's clarification immediately raised opposition, in particular from the secretary of the society, Robert Hooke. Newton moved on with equal success to other matters, and published a manuscript of his optical work under the title *Opticks* only in 1704, one year after the death of Hooke. It contains Newton's color circle demonstrating results of mixtures of spectral lights (see Fig. 5.1). Newton's work was admired, for example by Voltaire, but continued to be attacked during and after his lifetime. But he gained growing support from fellow scientists who recreated his experiments, thereby convincing themselves of their truth.

Painters and dyers were skeptical of Newton's findings. They knew by experience that all hues can be created (at greater or lesser saturation) with the use of three simple colorants only: yellow, red, and blue. Support for this idea was provided in (circa) 1725 by the efforts of the German painter and printer J. C. le Blon, who invented three-color printing by preparing separately engraved plates for yellow, red, and blue printing inks that, when printed on top of each other, resulted in multicolored images with beige, brown, and gray tones (17).

Newton explained the spectral composition of daylight and the color perceptions caused by its components, but did not address the color vision mechanism. The anatomical composition of the human eye was known in considerable detail at that time. In 1777 the English glassmaker George Palmer published a book, *Theory of Colours and Vision*, in which he proposed white light to be composed of three kinds of rays, resulting respectively, in yellow, red, and blue color perceptions (18). This was a step backwards from Newton, but he also suggested that there are three kinds of particles on the surface of the retina, each one sensitive to one of the three rays. Complete and uniform motion of the three particles produces the perception of white, the absence of motion that of black. Incomplete mixed motion results in the perception of chromatic colors. Interest in Palmer's proposals was slight and his book was soon forgotten. As a result, the English physician Thomas Young is usually credited with the idea of three different receptor types in the retina, proposed by him in 1802 before the Royal Society.

Color order had progressed by that time via two-dimensional color charts to three-dimensional systems. In 1771 the Austrian entomologist and Jesuit Ignaz Schiffermüller published a book *Versuch eines Farbensystems* (Attempt to construct a color system) in which he illustrated a twelve-grade hue circle, somewhat similar to one published in 1708 by an anonymous author in a French book on miniature painting (see Fig. 11.2) (19). Schiffermüller knew about tint/shade scales, but included only three scales of this type of blue colors from his circle. Around 1772 the English entomologist and engraver Moses Harris published a short illustrated text *A Natural System of Colours* with images of two circles, one titled "prismatic," the other "compound." The first contains eighteen hues at different intensity, that is, three primary and five intermediate hues between each of them. In the second the three secondary colors, orange, green and purple, are used to make fifteen intermediate mixtures. Harris, an experienced engraver, used black lines to impart blackness to his painted pigment colors (20).

In 1758 the German astronomer Tobias Mayer presented a public lecture in which he described a color system based on a mixture of three primary pigments, yellow, red, and blue, and filling a double triangular pyramid space. The written version was only published in 1775, after Mayer's death (21). A newspaper article about the lecture was read by the Swiss mathematician and astronomer Johann Heinrich Lambert, who engaged a painter to help him develop Mayer's ideas. Mayer had not proceeded beyond very simple coloration efforts for his proposed system and found the exercise very difficult. Similar problems arose for Lambert. He and his assistant Calau determined the relative coloristic strength of the selected primary pigments and mixed them in various ratios of two and three. Since they obtained black by mixing all three in a particular ratio, he did not see any need for the lower half of Mayer's double pyramid, supposed to be filled with different levels of black added to mixtures of primaries. His triangular pyramid has eight levels of increasing lightness, of which he showed only six in his book (Fig. 10.4). It represents the first somewhat systematic, illustrated, three-dimensional arrangement of color samples (22).

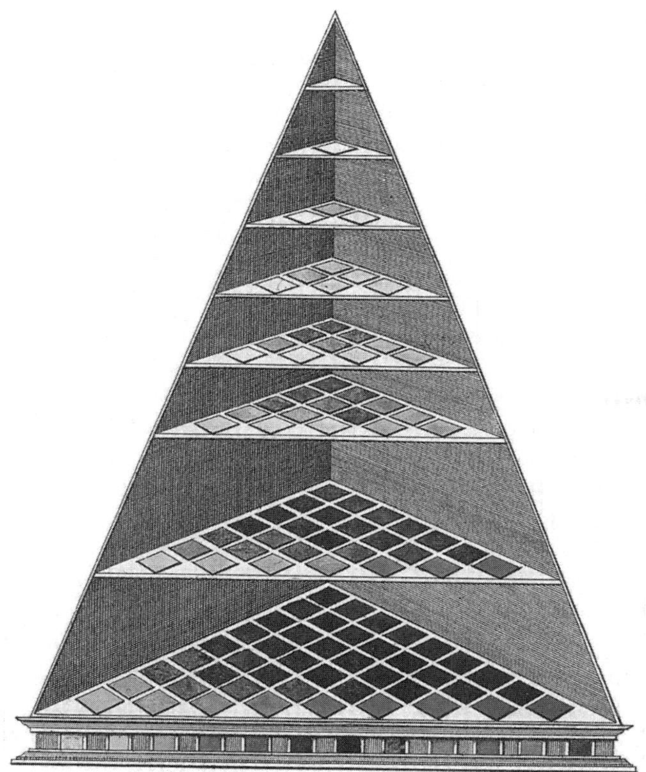

FIGURE 10.4 *Lambert's color pyramid, the first three-dimensional arrangement of color samples.*

A color system in the form of a sphere was published by the German painter Phillip Otto Runge (23) in 1810, the last year of his life. He developed it from a chromatic plane based on an equilateral triangle of primary and a second one of secondary colors. He arranged these colors on the equator of a sphere and placed a gray scale obliquely in the center, from white on top to black on the bottom. The remainder of the surface was colored with tint and shade scales from the equatorial colors to the white and black poles. Runge clearly understood that equivalent achromatic colors could be generated not only from mixing white and black pigments but also from mixing the chromatic primary pigments and mixtures in appropriate ratios. Opposing colors were taken to be compensative, neutralizing themselves in the center (Fig. 10.5).

An important contributor to discussions on color in the early nineteenth century was the German poet and natural scientist Goethe. Borrowing a prism from a friend, he concluded that Newton's findings were erroneous. By refracting not the light from a small beam of light, but as reflected from white paper with black lines, he had obtained, without realizing it, an arrangement inverse to that of Newton. In his *Farbenlehre* (Theory of colors) he sharply attacked the long-dead Newton and created a dichotomy of views on color that, in retrospect, was unnecessary. The work also includes a broad history of ideas on color, beginning with the ancient Greeks.

Unlike in the color systems of the past, the idea of placing colors of the same lightness on the same plane occurred at about the same time to a silk weaver in Paris and a painter in Munich. In about 1810, the French developer of a method to produce naturalistic images in silk, Gaspard Grégoire, published a color chart with 1350 color samples, and in approximately 1812, a book *Théorie des Couleurs* (Color theory) with a smaller color chart. The central plane of his cylindrical system has middle gray in the center of a twelve-step hue circle and seven steps of desaturated colors between the central gray and the saturated colors on the periphery. There are a total of five such constant lightness planes in the system. Then in 1816, the Bavarian court painter Mathias Klotz published *Gründliche Farbenlehre* (Thorough theory of colors), in which he illustrated what amounts to a 9-grade logarithmic gray scale and a circular constant lightness plane with a 24-grade hue circle and four saturation grades between the central gray. Only one of these planes was illustrated by Klotz. In terms of their structures (except for the open-ended chroma scale), these systems are prototypes of Munsell's system, which was developed 80 years later.

Simultaneous color contrast, long known, was investigated by the French chemist Michel-Eugène Chevreul. In his book *De la loi du contrast simultané des couleurs* (On the law of simultaneous color contrast), he introduced a hemispherical color space. By the middle of the century, the Austrian mathematician Christian Doppler, discoverer of the effect named after him, described a system of color classification in the form of a sphere octant, a form later also used by Erwin Schrödinger (24). In an 1831 book, *A Treatise on Optics*, the Scottish physicist David Brewster proposed that yellow, red, and blue were the primaries in the case of both light and object colors. He produced spectral curves purporting to show that yellow, red, and blue light exists in all colors of the spectrum. In this way, he believed he had solved the centuries old problem of how to reconcile the effects of light and object colors.

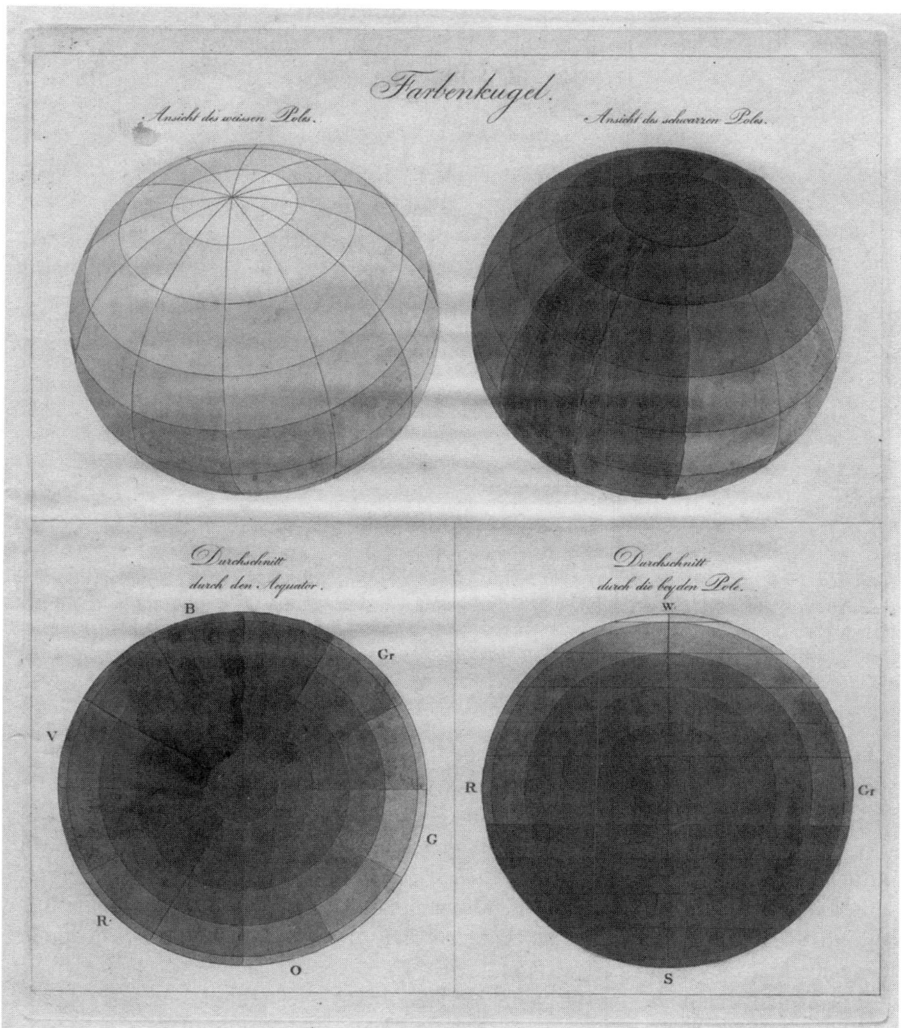

FIGURE 10.5 Phillip Otto Runge's color sphere of 1810. On top are views toward the poles. On the bottom left is an equatorial section, and on the right a vertical section along the central axis. (Image courtesy Werner Spillmann Collection.) Figure also appears in color figure section.

PHYSICS AND PSYCHOLOGY

By the middle of the nineteenth century, the German psychologists Weber and Fechner founded psychophysics, an attempt to quantitatively link the magnitude of sensory stimuli with the psychological experiences they engender. A perceptual distance was considered to be the sum of just noticeable differences between two percepts, say, two

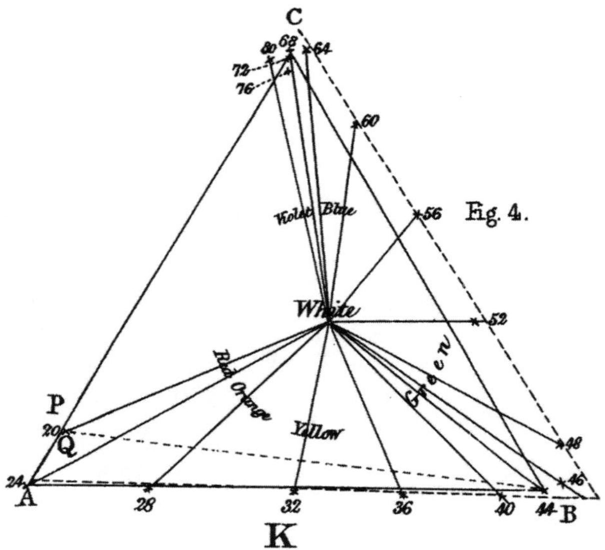

FIGURE 10.6 *Maxwell's color mixture triangle as experimentally determined for observer K. A represents the red, B the green, and C the violet spectral primary. The endpoints of the lines designated by (arbitrary) numbers represent the locations of the corresponding spectral lights in the diagram. The blank region on the left indicates the area of the nonspectral purple colors (26).*

sounds or two colors. Fechner determined the relationship between stimulus and response to be logarithmic. Later investigations by S.S. Stevens and others indicated that usually a power relationship applies. This is particularly valid for color scaling (25).

At the same time the English physicist James Clerk Maxwell, continuing work initiated by his erstwhile teacher Forbes, experimented with disk mixture and represented his results in what came to be known as the Maxwell triangle with the three primaries in the corners. Later he replaced disk mixture with a visual colorimeter using spectral rays as primaries (Fig. 10.6).

In Germany the physicist Hermann von Helmholtz investigated sounds and colors, and in an 1852 paper described in a comprehensive manner the difference between light and object colors. In the same paper he published a table with experimental results showing that only yellow and blue light when mixed formed white light, but not other combinations, as one should expect from Newton's color circle. The mathematician Hermann Günther Grassmann, having read Helmholtz's account, demonstrated, based on logic alone, that if Newton's theory was correct, any colored light could be matched with a combination of three primary lights. As a result, there must be many pairs of compensative lights that result in white light, if appropriately mixed. Helmholtz repeated his experiments, and less than a year later published a paper confirming Grassmann's conjectures and Newton's views. Grassmann also postulated three laws that became fundamental for trichromatic color theory (see Notes in Chapter 6). Using a visual colorimeter of Helmholtz's design, his assistant

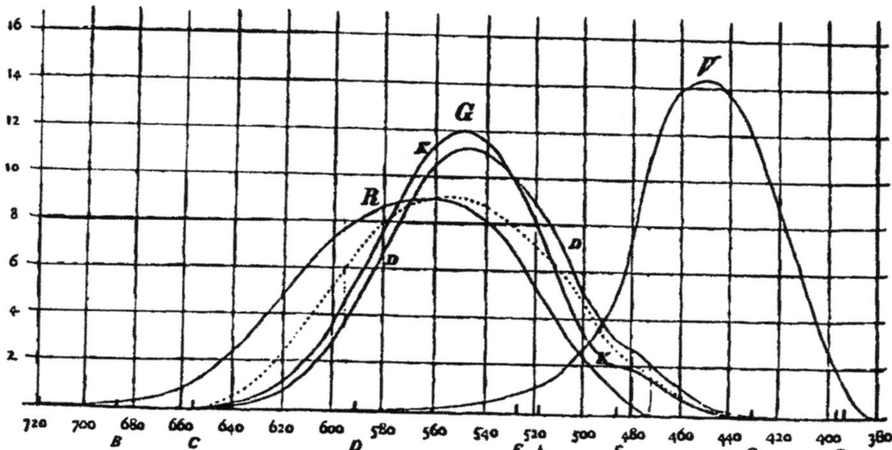

FIGURE 10.7 *The fundamental sensitivity curves as measured by Artur König and his assistant Dieterici. Note the reversed wavelength scale.* R *represents the red,* G *the green, and* V *the violet fundamentals. Separate* G *curves for König and Dietrici are shown. The dotted curve represents the sensitivity of visual purple (rod vision).*

Artur König made accurate measurements of the three fundamental sensations, the sensitivities of the three cone types (Fig. 10.7). The measurements indicated that these sensitivities can vary significantly among observers and that certain observers with impaired color vision lacked one or more of the sensitivities. The new, experimentally supported, theory of color vision, published in Helmholtz's *Handbuch der physiologischen Optik* (Treatise on physiological optics) of 1867, came to be known as the Young–Helmholtz theory. Helmholtz defined three attributes for color: hue, saturation, and brightness or lightness. They had been suggested in principle already by Newton, but Helmholtz defined them based on stimulus information (27).

Almost immediately Helmholtz began to face serious criticism from the physiologist Ewald Hering. In the mid-1870s, the latter proposed a theory based on an idea of the physiologist Hermann Aubert, who believed that black, white, red, yellow, green, and blue are the principal color perceptions. Hering concluded that there are three antagonistic or opposing pairs, white and black, yellow and blue, and red and green, and all other color perceptions are composed of these basic perceptions. He believed the corresponding physiological processes to be located in the retina. The chromatic *Urfarben* (fundamental colors) are pure. For example, fundamental blue is the blue of highest intensity that is neither greenish nor reddish. His ideas are supported by the psychological fact that there is no yellowish blue and no reddish green, thus these pairs oppose each other. The situation is different in the case of black and white, as their mixtures form a series of grays. Hering proposed that any perceived color is a mixture of perceptions of one or two chromatic fundamental colors, black, and white, and he represented them in an equilateral triangle (28).

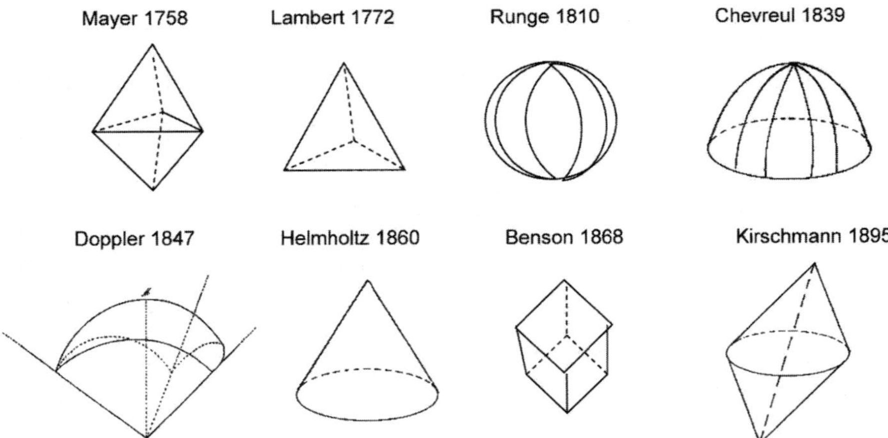

FIGURE 10.8 *Schematic representations of geometrical solids proposed as color solids by various authors.*

Helmholtz recognized that the two theories do not have to be mutually exclusive. The physiologists Donders and von Kries separately proposed a combination of the two theories in a "zone" theory where the Young–Helmholtz theory applies at the cone level and the Hering theory at a later step in the visual process. However, the Young–Helmholtz theory was the leading paradigm well into the twentieth century (29).

Psychological research in color progressed and resulted in several different proposals for geometrical color solids: a cone by Helmholtz, a tilted cube by Benson, a square double pyramid by Höfler, a tilted double cone by Kirschmann, a tilted double pyramid by Ebbinghaus and Titchener (Fig. 10.8) (30).

Many perceptual color phenomena, such as brightness and lightness perception and color adaptation, were investigated, and in the latter case von Kries developed a simple mathematical procedure for (approximately) predicting the results of adaptation. The measurement of the brightness of light had been a subject of interest since the seventeenth century, and in 1760 Lambert wrote a book about it where he predicted that in the future it would be possible to measure the brightness of a light with an objective instrument, like temperature (31). In the meantime, candles or other standard lamps were used as reference. Visual photometers were developed in the nineteenth century, and late in that century light-sensitive photoelectric devices were invented, making objective measurement possible. It was discovered that not all spectral wavelengths contribute equally to the perception of brightness. In 1828 the German physiologist Treviranus discovered the separate nature of rods and cones in the retina, and Boll discovered visual purple in the rods. A "duplicity" theory of vision, according to which rods are responsible for night and cones for day vision, was proposed in 1866 by the German anatomist Schultze and then expanded by von Kries. Later König found close experimental agreement between the sensitivity curve of a color-blind (but not blind) person, that of a color normal-observer at very low intensity, and the absorption curve for visual purple (retinal), thus confirming the validity of the duplicity theory (see Fig. 10.7) (32).

COLOR ORDER IN THE TWENTIETH CENTURY

Color order continued to make progress also. In the first years of the twentieth century, the American painter and educator Albert Munsell conceived a balanced color sphere based on the three attributes hue, value (for lightness), and chroma (for intensity). He based it on five primary colors, since he did not believe Hering's theory to be valid and because he favored the decimal system. Within each attribute the perceptual distances between grades were to be uniform. From his efforts to color samples in series according to attributes, Munsell quickly learned that the maximum chroma achievable with different pigments varied significantly. He made the chroma scale open-ended, and thereby had to give up the sphere form of his system. Munsell published a first atlas in 1907, followed by an expanded one with 880 samples in 1915, and the first version of the *Munsell Book of Colors* was published in 1929. The American ornithologist Robert Ridgway, who was interested in defining bird colors, asked physicists for help in designing a system based on disk mixture, and in 1912 he self-published a color atlas with 1115 chips under the name *Color Standards and Color Nomenclature*.

In 1917 the Nobel Prize–winning German chemist Wilhelm Ostwald began to publish a series of books on color under the title *Die Farbenlehre*, in which he described a double cone color solid based on three primary colors and the concept of *Vollfarben* (full colors, in the ideal case samples made with (nonexisting) pigments with sharp transitions in reflectance between 0 and 100%). The samples were ordered, as in Hering's system, according to their content of full color, blackness, and whiteness. To obtain what he thought was perceptual uniformity, Ostwald spaced his samples according to the Weber–Fechner law of psychophysics. Ostwald's *Farbenatlas* of 1917 contains 2500 samples, the first extensive, systematic arrangement of object colors. Ostwald's system was influential in much of Europe until World War II. It was reproduced in the U.S. under the name *Color Harmony Manual*. Ostwald also was the first to define the concept of metamerism in terms of reflectance functions.

With the developing capabilities of the measurement of the spectral power of lights and the reflectance of objects color became increasingly equated with the stimulus despite the fact that perceived color does not stand in a simple relationship to the stimulus, as some psychologists continued to point out. Mathematization of color theory, beginning with Newton, increased apace. In 1920, Erwin Schrödinger, the Austrian co-founder of quantum mechanics and author of the influential book *What is Life*?, developed a mathematical theory of color stimuli, laying the foundation for much of the later development. Calling it "basic color measurement," he developed the trichromatic theory and defined the shape of the spectral envelope (Schrödinger's spectrum bag; Fig. 10.9) and the optimal object color solid based on ideal color stimuli

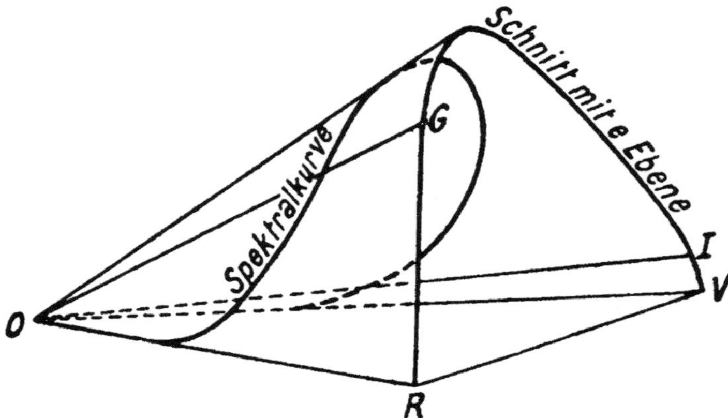

FIGURE 10.9 *Erwin Schrödinger's* Farbentüte *(spectrum bag). O is the black point. The primary sensations* R, G, *and* V *represent red, green, and violet. The trace of spectral colors is curved on the surface. The straight line from red to violet represents purple colors not in the spectrum. Object colors form a solid inside the bag.*

as earlier defined by Ostwald. Under the designation "advanced color measurement," Schrödinger mathematically defined what became known as color metrics, the internal spacing of the ideal Euclidean object color stimulus solid, so that colors were perceptually uniformly spaced. As seen in Chapters 5 and 7, this problem turned out to be much more complex than originally thought (33).

In the early 1920s the CIE was formed, and in 1924, based on new measurements, it defined the standard photopic luminosity curve, expressing brightness perception as a function of spectral wavelength. In 1931, it recommended standards and a numerical system for technological implementation of the trichromatic system. Using the spectral sensitivity curves of a standard observer, the spectral power of three standard light sources, and methods for measurement of spectral reflectance and transmittance, three tristimulus values and three chromaticity coordinates were defined, making possible the identification of a color stimulus with three numbers (see Chapter 6).

The Optical Society of America redefined the Munsell system in 1943, and completed development of a more generally (approximately) uniform system (OSA-UCS; see Chapter 5) in the 1970s. The first color difference formula, based on the Munsell system, the Nickerson Index of Fading, was constructed in 1936. Judd, Godlove, and MacAdam (and many others) worked on the mathematical formulation of uniform color space, a process that is not yet complete. The CIE recommended several color space and difference formulas, among them CIELAB and CIELUV (see Chapter 6).

Hering's ideas of opponent colors received support in the 1960s from hue cancellation experiments by Hurvich and Jameson. As mentioned in Chapter 3, cells with opponent character have been discovered in the retina and lateral geniculate nuclei in the brain. As a result, all modern color space and difference as well as appearance formulas have a zone structure with one step involving an opponent color system.

COLOR TECHNOLOGY AND COLOR SCIENCE

CIE recommendations for mathematical description of various perceptual effects based on spectral sensitivity functions, such as color rendering of light sources and the changes in perceived color resulting from adaptation to different light sources, were also developed. In the absence of detailed knowledge about the neurological operations behind these effects, such formulas must be regarded as exercises in mathematical fitting of perceptual data.

In 1932, the German physicists Kubelka and Munk published an analysis of the relationship between light absorption and light scattering by reflecting objects and the resulting reflectance function (see Chapter 9). The results provided a basis in the 1950s and later for the development of optimized colorant formulation (see Chapter 9). Rapid technological development in computer-assisted reflectance instrumentation and minicomputers advanced this technology to a point where today it can be found in paint stores.

On the anatomical front, dramatic progress was made in the twentieth century in the analysis of the visual pathway in the brain. The path of the optic nerve from the eye to the main visual processing center in the back of the brain is now quite well understood, both anatomically and physiologically (34). However, this has not resulted in a full understanding of the processes of vision as yet. Mathematical models of the complete color vision process have been attempted several times, for example, by E. G. Müller, as expressed in mathematical form by Judd, the ADT model by Guth, and the multistage model by De Valois (35). They either require special fine-tuning for different perceptual effects or predict some well and others poorly, indicating the lack of detailed knowledge about such effects.

Sensory perceptions, such as vision and color, have proved to be very difficult to understand. They are wrapped up in the riddle of consciousness, as discussed in Chapter 2. Just as older authoritative ideas like Aristotle's seven simple colors, or Brewster's common primaries of yellow, red, and blue for light and object colors, have proven wrong or incomplete, it is likely that much of what we consider valid today will require additions or replacement in the future. The color vision mechanism is very likely much more complex than today's simple, cone-based models imply. Human variation in color vision appears to have considerably more variability than believed up to now on basis of a rather limited number of color-matching function measurements. Our neural nets active in object recognition have likely developed to some extent differently in different groups due to the ways variation in long-term visual experiences of early ancestors affected their tuning. Humanity's journey in understanding color vision, very briefly sketched in the preceding paragraphs, will continue for an unforeseeable time.

11

Color (Theory) in Art

The question of the essence of pictorial art and beauty has been the subject of thought and discussion since ancient times. The philosopher Plato wrote that the result of drawing and painting is "dreams created by man for those who are awake." He believed the purpose of pictorial art to be imitative, but image making could be either imitative or represent imagination or the fantastic.

Two thousand years later Kant described the beautiful to be "that which pleases universally without requiring a concept." He thought that to produce beautiful art required genius, products of beautiful art are a combination of taste with genius. Beautiful art pleases, but it also raises feelings of the sublime (1).

Thinking about the purpose and meaning of pictorial art in human life has continued unabated, but the subject is found too complex to have its essence caught in a few sentences. More recently, creation of pictorial art has become a subject of interest to neurophysiologists. Ramachandran and Hirstein, on the basis of ideas of neural processing, proposed eight principles, "eight laws of artistic experience." These are the peak-shift principle (essentially exaggeration resulting in caricature), isolation of a single cue, perceptual grouping to delineate figures, extraction of contrast, perceptual problem solving, abhorrence of unique vantage points, use of visual metaphors, and symmetry (2). Several commentators found this reductionist approach unsatisfactory for explaining the often deep and rich emotional experiences of viewing pictorial art. But others agreed that with this contribution new ground has been broken in the age-old discussion about art. Insofar as the creation and experience of art is a human activity, it has to have its maps and activities in the brain. Another neurophysiologist, Zeki, believes it to be no coincidence that the ability of artists to abstract essential

Color: *An Introduction to Practice and Principles, Second Edition*, by Rolf G. Kuehni
ISBN 0471-66006-X Copyright © 2005 John Wiley & Sons, Inc.

features of a visual image by discarding redundant information is essentially identical to such activity by the visual system. Zeki studied the activity in various areas of the brain when viewing natural objects or various kinds of pictorial art with brain scans and found distinct differences, particularly for art that involves no relationship to natural objects or scenes (3).

The range of pictorial art is huge, from the images in the caves of Southern France and Northern Spain, the paintings and frescoes of the Renaissance, the fantastic images of Blake, to realistic color photography, the huge canvases of abstract-expressionist paintings, and the drip canvases of Pollock. Such images create vastly different impressions in different viewers as a result, presumably, of their brain's wiring and their personal cultural and experience history. If there is a common essence to what we consider beautiful and/or sublime, it remains to be understood.

As we now know, our ancestors expressed themselves in lines, colors, and volumes as early as 35,000 years ago. A spectacular example, whose oldest artifacts range back nearly that far, is the Chauvet Cave (see Chapter 10). Line drawings and paintings with shading and highlights in black, white, and various ochre shades depict a range of different animals as well as humans, many of surprising levels of naturalism, others more fantastical and mystical. The impulse can only be speculated about: an urge to fix certain facts of life, a plea for the continuation of success in the all-important hunt and battle with enemies, a need to free the memory from the terrors of hunt and battle by depicting them on the wall. Verbal language was probably in the earliest stages at that time and writing more than 25,000 years in the future. While it is generally assumed that brain organization was largely identical to today's, there is little doubt that the mental structures and contents must have been vastly different. Recently the close resemblance between such cave art and sketches by some modern autistic children of early age has been pointed out, not so much to claim that our ancestors at that time were autistic, but that their conscious brain processes were not dominated by the verbal to the extent they are today (4).

By the end of the Ice Age, some 10,000 years ago, human tool use had advanced significantly, with bows and arrows, spear throwers, and harpoons in wide use. Some of these were made more distinctive by the addition of pictorial elements, mostly animal images. Funerals had become common and caves often contained artifacts testifying to the importance of, reverence for, or love for the deceased. The exclusively nomadic life as hunters and gatherers began to draw to a close in certain regions. There, wild grain was collected, skin-covered huts began to be built, and lamps were in use.

Farming began to spread from the Middle East north, east, and westward some 8–10,000 years ago. The development of farming made possible a degree of specialization of activity. Skillful toolmakers and artisans were able to trade their work for food. Such activities resulted in the joining of clans of people into villages and later towns. Pottery appeared, first sun-dried and unornamented, later kiln-fired and more and more and more lavishly decorated. By 5000 B.C., polychrome pottery had been invented and appeared in Mesopotamia and Turkey. Wall painting in and on houses depicting hunting scenes, images of a goddess cult, and of an erupting volcano are known from the same period in the examples in the Turkish settlement of Catal Höyük.

Between 5000 and 3000 B.C., a veritable explosion of technology took place. Copper was discovered, probably in the eastern Mediterrenean region, resulting in advances in weaponry, tools, and adornments. The wheel and writing in pictographs appeared in Mesopotamia. Egyptians invented weaving, perhaps glass, and their own form of pictographic writing, hieroglyphs. Also in Mesopotamia, an early form of mosaic was created by embedding short colored pottery rods in plaster.

By 3000 B.C. knowledge of how to produce polychrome pottery had expanded over a wide area, from England in the West to China in the East. It represented the most common and technically most advanced medium of polychromatic art. Between 3000 and 1500 B.C., the greatest advances in all areas of civilization appear to have occurred in Mesopotamia, Egypt, and China. The first known literary works, the *Epic of Gilgamesh* in Mesopotamia and the *Story of Sinhue* in Egypt, appeared. Metallurgy of several metals, including silver and gold, was known. Coloration technology improved with the discovery of grindable semiprecious stones (lapis lazuli, turquoise, azurite, malachite) and of artificially produced pigments. Pottery and enamel reached new levels and colored glass became ubiquitous. The state of coloration technology of murals, statues, jewelry, leather goods, textiles, wood, and other materials can be inferred, for example, from some of the extensively painted grave chambers and artifacts found in Egypt's Valleys of the Kings and the Queens. Knowledge of such technologies spread north and westward to the Minoan palaces in Crete, to Greece, and further west. On the American continent, beginning as early as 20,000 B.C., immigrants from Northeastern Asia, traveling via the Bering Strait into Alaska and down the continent, developed their own technologies. Textiles dyed with plant dyes and alumn mordant and dating back to ca. 1000 B.C. have survived in current Peru. Chinese and Indian cultures began to flourish in the third and second millenia B.C., producing polychromatic pottery and textiles.

By 500 B.C. Egyptian culture began to decline, Mesopotamia had reached its apex, and Hellenistic culture advanced rapidly. In the Americas local cultures began to build pyramids and produce polychrome pottery. The next 500 years saw the spread of Greek culture and the development of the Alexandrian and Roman empires. Iron replaced copper and bronze as the metal of choice for weapons and tools. Polychrome painting flourished in Greek temples and houses. Mosaics reached high levels of perfection. Colorful textiles from this time period have been preserved in regions where climatic conditions are favorable. Felt appliqué hangings and mummy wrappings have been found in Siberia, Mexico, and Peru.

In Greece, early philosophers developed the first theories of color and esthetics. Plato's and Aristotle's ideas had a far-reaching impact and consequences. As mentioned in Chapter 10, Aristotle posited a list of seven simple colors: white, yellow, crimson, violet, leek-green, dark blue, and black, the five chromatic colors themselves believed to have been generated in some fashion from white and black. This view influenced European thinking about colors into the seventeenth century.

In the first century A.D., Pliny the Elder wrote a natural history, including a canon of great artists of the Greek classical world. According to Pliny, shading was introduced by the painter and sculptor Apollodorus, and the use of highlights in addition to shading by Zeuxis in the fourth century B.C. The most highly regarded Greek painter was Apelles, court painter to Alexander the Great. The classical range of colors was, according to Pliny, limited to white, yellow, red, and black, posited by the Pythagoreans. The famous mosaic of Alexander in battle is limited to these four colors and may represent a typical example in terms of colors. Apelles is reported to have protected his paintings with a glossy finish containing a small amount of burnt ivory black, thereby reducing the high chroma of yellow and red pigments and resulting in a preferred subdued appearance. The absence of blue and green is surprising. There is no doubt, however, that in Apelles's time Greek painters knew indigo as a pigment, as well as azurite and lapis lazuli. Blue pigment traces have been found with others on temple friezes, but perhaps such materials were not used in panel painting until later. Pliny lists some twenty-two different pigments in his book, including natural products, whites, ochres, malachite and indigo, as well as manipulated or manufactured products, for example, orpiment, realgar, minium, purple, and verdigris (5).

In the second and first centuries B.C., several Greek painters moved to Rome and thereby transferred styles and material knowledge. In the Roman empire, panel painting slowly faded in favor of wall paintings. The number of pigments in use grew steadily, as attested by Pliny. The most extensive collections of examples of Roman painting are found in the ruins of Pompeii and Herculaneum, both destroyed by the eruption of Mount Vesuvius. Image subjects range from the purely decorative to complex allegorical paintings and landscapes. A most impressive example is a series of paintings in the Villa of the Mysteries in Pompeii, showing the initiation of a young girl into the rites of Dionysius. During the decline of the Roman empire much artwork was destroyed, but a few examples have been found in the provinces, indicating insignificant advances. Miniature paintings on manuscripts are found for the first time in the fifth century A.D.

Artistic achievements comparable to those in Europe from the same time period have been found in China, India, and Mesoamerica, primarily in grave chambers. A rising Christianity produced its first strongly polychromatic works of art in the mosaics of early churches in Rome (fourth and fifth centuries A.D.). For the next 700 years European polychrome art was primarily commissioned by the Christian church, in the form of church decorations and manuscript illumination. The *Lindisfarne Gospels* (seventh century) and the *Book of Kells* (ninth century) are prime examples of the latter, the Hagia Sophia church in Istanbul (ninth century) and San Marco church in Venice (eleventh century) of the former. Particularly beautiful examples of mosaic work from the sixth century are found in the Adriatic coast city of Ravenna.

The twelfth and thirteenth centuries saw the building of great Gothic cathedrals in France and elsewhere in Northern Europe. When constructing the cathedral of St.-Denis in Paris in the first half of the twelfth century, the responsible abbot Suger decided on using stained-glass windows as one of its major chromatic decorations. Coloration of glass is an art known at that time since at least 2000 years, but Suger and his craftspeople developed entirely new applications in cathedral windows. Stained-glass windows became the standard for Gothic cathedrals for the next two centuries, and prime examples of the art are found in the cathedrals of Notre Dame in Paris and in Reims (6).

Manuscripts on colorant manufacture and application date back to Egyptian papyri. This kind of information was handed down through history and regularly enlarged and updated. Among the early manuscripts is the twelfth century *Mappae clavicula*, itself a compilation of older information. In 1431 the French lawyer Jehan le Begue compiled several earlier manuscripts, among them one attributed to an Italian monk named Eraclius (7). Book I of Theophilus's twelfth century treatise on the practical arts of church adornment, *De diversis artibus* (The various arts), is titled "On the temperaments of colors" and describes what colors and pigments, and in what forms, to use for painting specific details. Book II discusses the manufacture and use of stained glass, and Book III discusses gilding and metalworking. In Book I, he describes how to create multiple levels of shading and highlighting in painting, resulting in a twelve-grade scale, as mentioned in Chapter 10.

Frescoes and panel painting began to replace mosaics in church decoration in the thirteenth century. In Italy Cimabue and Giotto were early masters of panel painting.

THE RENAISSANCE

Beginning in the fourteenth century, the political climate in Italy and elsewhere in Europe was dominated by competing nation states with generally strongly autocratic leaders who often supported the arts. This resulted in a rediscovery of the classical poets and philosophers and integration of certain aspects of classicism into Christian thinking. It also resulted in a renewed interest in nature and efforts to discover its secrets. Around the year 1400, the architect Brunelleschi discovered the rules of perspective, of particular importance to painting. The correct treatment of perspective rapidly became important in panel painting, and knowledgeable artists guarded its secrets. At about the same time, the custom of painting large fresco cycles in churches and princely palaces began. Such impulses resulted in an outpouring of polychromatic art that is perhaps unparalleled. A few examples of fresco art of the early Italian Renaissance are those by Masolino, Masaccio and Lippi in the Brancacci chapel of Santa Maria del Carmine in Florence, Fra Angelico in the Chapel of Nicholas V in the Vatican, or Gozzoli in the Chapel of the Magi in the Palazzo Medici-Ricardi in Florence. The artists of the fresco also produced panel paintings of startling chromatic inventiveness. Most of the work involved religious themes, but the practice of portraiture, nature painting, and painting of mythological themes (for example, Botticelli's *Birth of Venus*) also became common.

Theories of color to support the work of the painters began to appear and assume some importance. But, unsurprisingly, a single theory never managed to become generally accepted. Plato's and Aristotle's ideas on the nature and meaning of color, passed on through the Middle Ages by Galen (second century), and the work of the Arabic philosophers Al Kindi (ninth century), Avicenna (tenth century), and Averroës (twelfth century), received new interest, scrutiny, and interpretation. Theories of painting involving many other aspects in addition to color were developed by Cennini (*Libro dell'Arte*, The Craftsman's Handbook, ca 1390), Alberti (*Della Pittura*, On Painting, 1435), Piero della Francesca (*De Prospectiva Pingendi*, On Perspective in Painting, 1482), Leonardo da Vinci (Notes toward a *Trattato della Pittura*, Treatise on Painting, ca. 1500), and Lomazzo (*Trattato dell'Arte de la Pittura, Scoltura, et Architettura*,

Treatise on the Art of Painting, Sculpture, and Architecture, 1584, and *Idea della Pittura* The Idea of Painting, 1590). In 1550 Vasari published his description of the lives of the eminent architects, sculptors, and painters of the Italian Renaissance (*Vite de' Più Eccelenti Architetti, Pittori ed Scultori Italiani*, Lives of the most excellent Italian Architects, Painters, and Sculptors, 1550 (8). A compilation of knowledge and ideas about painting and color was published in 1584 as *Il Riposo* (The Repository) by Borghini (8).

Cennini was a painter himself, and his practical experience is obvious. He described a drapery modeling technique that became known under his name. The pure pigment is used to paint the darkest area of drapery and all modeling is achieved with increasing dilutions of the chromatic pigment with white pigment. As another possibility for drapery modeling, he mentioned the use of *cangiante* (see later). In contrast Alberti recommended modeling in both directions toward white and black, as was suggested earlier by Theophilus.

The Italian theorists decided that a painting required beauty (*belleza*), design (*disegna*), a clear relationship between light and dark (*chiaroscuro*), and color. Excellence of paintings (*paragone*) was widely discussed and aspects such as nobility, utility, truth, difficulty, and ornamentation were considered important. Of particular importance was the natural presentation of skin complexion (*incarnazione*). Lifelikeness (*vivacità*) was highly praised, as was the achievement of the illusion of a third dimension (*rilievo*). A painting did not just require accurate spatial perspective but a correct color perspective; the change in coloration from local to distant. To imbue images of certain persons with an aura of the sublime, some artists clothed them in fabrics displaying the *cangiante* effect (a silken fabric with contrasting colors of warp and weft, resulting in changes in color appearance as a function of the viewing angle; prime examples are found in Michelangelo's work on the ceiling of the Sistine chapel).

A painting started out with a concept or idea. Its realization required manual and material obedience, and the final work needed to have durability. The selection of color involved harmony and ideas about their deeper meaning: a symbolic content. Such ideas had been discussed by Plato and Aristotle already, and received modifications due to the usage of color in churches and for other reasons. Treatises discussing these ideas were published in 1565 by Dolce (*Dialogo di M. Lodovico Dolce nel quale si ragiona delle qualità, diversità e proprietà dei colori*, Dolce's dialog on the quality, diversity, and appropriateness of colors), in 1568 by Occolti (*Trattato de Colori*), and in 1595 by Calli (*Discorso de'colori*, Discourse on colors) (9). Arguments about the symbolic content of colors remained unresolved, however. Colors have been equated to the four elements of antiquity, for example, by Alberti: red–fire, blue–air, green–water, beige or yellow–earth; to the four human temperaments, for example, by Borghini: blue–sanguine, red–choleric, dark violet–melancholic, white–phlegmatic; to celestial bodies, for example, Borghini: yellow–sun, white–moon, red–Mars, blue–Jupiter, black–Saturn, green–Venus, purple–Mercury; to musical notes, for example, by Aristotle or Arcimboldo.

In the sixteenth century, Western polychromatic painting reached lofty heights in the works of Michelangelo, Raphael, Leonardo, Titian, and others. On the northern side of the Alps, German, Dutch, Flemish, and French painters and illustrators, such as the Limbourg brothers, van Eyck, van der Weyden, Grünewald, Bosch, the Brueghels, created works of comparably high value, but imbued with their own cultural traditions. In Flanders and France, weavers produced highly sophisticated polychromatic tapestries by copying painting techniques (for example, the Unicorn Cycle).

FROM THE SEVENTEENTH TO THE NINETEENTH CENTURIES

Italian painting is often thought to have declined in the seventeenth century into the baroque, the highest achievements shifting north and west, as exemplified in the works of Rubens, Vermeer, van Dyck, Velàzquez, Hals, Rembrandt, and Poussin. The seventeenth century was the time of major developments in many areas. European nations competed for worldwide trading empires. Science began to break away from the classical dogmas about the world and humans that held sway over centuries, as exemplified by the new theories of astronomers (Galilei and Kepler) or anatomists (Harvey). Philosophers Descartes and Spinoza used the haven of Holland to offer new world views that challenged church doctrines. Grimaldi in Italy and Newton in England investigated the refraction of sunlight by a glass prism with far-reaching impact on color theory.

The latter efforts were in considerable contrast to the conventional views on color offered in mid-century by d'Aguilon and Kircher, both still representing the scholastic view of five simple colors (reduced from Aristotle's seven, see Chapter 10). In the seventeenth century, two key proponents of differing views about the application of color in painting were Peter Paul Rubens in Holland and Nicolas Poussin in France (and later in Rome). Poussin painted classically composed allegorical and mythological scenes, such as "Landscape with Orpheus and Eurydice," as well as religious themes, for example "The Judgment of Solomon." Scenes of human upheaval are set in calm, classical landscapes painted in naturalistic stile. In some of the paintings quite highly saturated colors favoring the painter's primaries yellow, red, and blue are used to designate the rank and status of some people as well as to create contrasting highlights that attract the eyes to certain areas in the painting. Poussin believed that paintings should not be lifelike reproductions of all the accidental details found in nature, but instead should represent a conscious design with a theme. In a self-portrait of 1659, he is seen holding a book titled *De Lumine et Colore* (On light and color), attesting to his interest in color. A book with this title is not extant today and it is not known who its author may have been (10). Poussin is likely to have known Kircher, as both were living in Rome at the same time. Poussin received technical inspiration on perspective in general and color perspective in particular from his acquaintance with Matteo Zaccolini, a Roman painter and author of a four-volume work on color and perspective published between 1618 and 1622.

In Holland Rubens pursued similar subject matter as Poussin, but in a much different style of painting. Some of his paintings and color sketches can be seen as impressionistic in nature. Rubens was acquainted with d'Aguilon, the author of

FIGURE 11.1 *One of six allegorical images designed by Peter Paul Rubens for d'Aguilon's* OpticorumLibriSex. *This image shows a photometric measurement. The distance of a single-flame lamp from the screen of observation is about half that of a two-flame lamp, resulting in equal perceptual brightness of the fields on the screen.*

Opticorum Libri Sex (Six books on optics), an influential book on optics. Rubens may have contributed to its text, and he designed its title page and an allegorical illustration for each of its six chapters (Fig. 11.1). He is known to have kept notebooks (now lost) believed to have contained his ideas on color and painting in general. D'Aguilon described three kinds of color mixture: mixture of colorants, mixture of intentional colors, as when an object is reflected from the shiny surface of another object with a different color, and notional mixture when "patches small enough to escape the eye" combine so that "for each of the combinations of colors a uniform color is received." Rubens made broad use of the last technique of optical mixture by providing texture and lifelike appearance, particularly in renditions of human flesh, with yellow, red, blue, green, and gray patches and streaks. Rubens created at least one painting with color as its major theme: *Juno and Argus*, 1611. It depicts Argus, the son of Zeus and Niobe, slain by Hermes. Argus's head was covered with many eyes and he was said to be all-seeing (Panoptes). In Rubens's painting, Juno transfers the eyes from Argus's head to the feathers of a peacock. The scene is illuminated by the sun forming a rainbow against dark clouds in the background (11). The different viewpoints on color expressed by the two artists were soon recognized as antithetical by commentators, who split into two groups: the Poussinists and the Rubenists. A chief Rubenist was Roger de Piles, a French painter and color theorist, who in 1684 wrote an introductory text for students of painting, *Les Premiers Élémens de la Peinture*

Pratique (First elements of practical painting), in which he suggested a palette with eight colors. In his *Dialogue sur le Coloris* (Dialog on coloration) of 1672, he bemoaned the fact that drawing and perspective had a scientific basis and was taught well in art schools, but color was lacking in both areas.

In the second half of the seventeenth century, Vermeer produced his iconic paintings of life of the Dutch bourgeoisie, with a predominance of yellow and ultramarine blue. He is now believed to have painted some of his canvases from images on the screen of a camera obscura.

In 1672 the young Newton described his findings on the nature of sunlight and thereby of color to the Royal Society in London and raised a storm of criticism lasting for over one hundred years and resulting in confusion by lay people. Are there three or five primary colors, or seven as Newton claimed? In 1720 Le Blon seemingly answered the question by demonstrating a three-color reproduction process in mezzotint printing, initiating efficient image reproduction in color (as noted in Chapter 10).

As the previous history shows, books on the craft of painting were written quite regularly. In Paris in 1672 a how-to book with the title *Ecole de la Mignature* (School of miniature painting) was published without the name of an author. Miniature painting was a specialized subset of panel painting since the beginning of manuscript illumination in the Middle Ages. The book is attributed to one Claude Boutet. There were at least 33 editions over the next 200 years, and translations into several languages. The book was also published in French by the Dutch publisher van Dole who, in his second edition of 1708, named *Traité de la Peinture en Mignature*, added new chapters and an addendum on pastel painting. Their author(s) is (are) unknown. As the publisher states in the section on pastel painting "... one finds there something quite curious concerning the primitive colors and the generation of composite colors. ..." The primitive colors are described as yellow, fire red, carmine red, and blue. Neutral red is mixed from the two primitives. Yellow and fire red make orange, carmine red and blue make violet, and blue and yellow make green. Mixture of primaries and secondaries produces additional hues, such as purple from carmine red and violet. Here, two hand-colored printed charts with images of color circles are included (see Fig. 11.2). They are described as follows: "Here are two figures that show how the primitive colors, yellow, red, carmine, and blue generate the other colors, one could name it the Encyclopedia of Colors. The first figure contains the four primitive colors with their three composites and the second contains the same colors with an additional five, in part from primitives and in part from composed" (12). These are the first known complete color circles appearing in print.

Of considerable influence on painters and students of art at the time was the book *Het Groot Schilderboek* (The art of painting, 1707) by the Dutch painter and theorist Gerard de Lairesse, translated into several languages and receiving several editions. In 1762 the German painter Mengs, considered preeminent at his time, wrote a brief but influential text, *Gedanken über die Schönheit und den Geschmack in der Malerey* (Reflections on beauty and taste in painting).

Goethe's occupation with matters of color was largely caused by a desire to help his artist friends with a system of color harmony and systematic application of colors to represent certain ethical values. At the time of writing his *Farbenlehre* (Color theory,

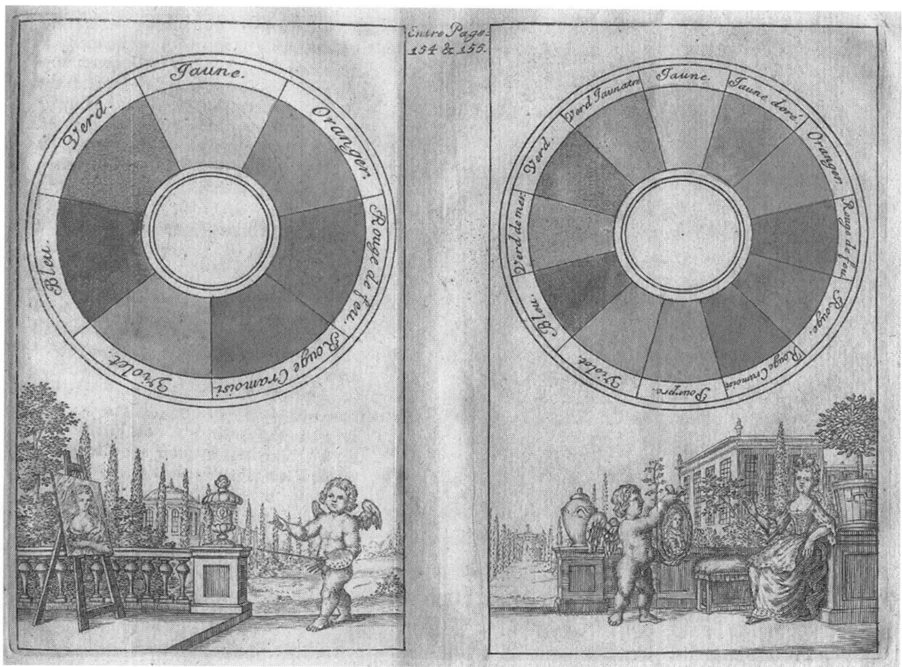

FIGURE 11.2 *The hand-painted color circles from the van Dole edition of 1708 of* Traité de la Peinture en Mignature. *(Image courtesy Werner Spillmann Collection.)* Figure also appears in color figure section.

1810), Goethe befriended the young painter Runge, the author of *Farben-Kugel* (Color sphere) (see Chapter 10 and Fig. 10.5). Runge often used complementary colors in his paintings, as well as colored shadows.

J. M. W. Turner, an English painter of the nineteenth century, had considerable interest in color in general and Goethe's theory in particular. In his later works, Turner had a preimpressionistic style, exemplified by his well-known painting *The Fighting 'Temeraire'* (1839). Turner also painted *Light and Colour (Goethe's Theory)–The Morning After the Deluge–Moses Writing the Book of Genesis*, in which light and colors have returned after 40 days of rain (13).

The main battle between old and new views of color in painting in the nineteenth century was fought between the classicist Ingres and the preimpressionist Delacroix. They were perhaps the logical outcome of the Poussinist–Rubenist division. Each denigrated the work of the other. Ingres's paintings have the classical outlines of figures and coloration that harks back to Renaissance symbolism and Poussin. Delacroix wrote about him: "The livid and leaden tones of an old wall by Rembrandt are far richer than this abundance of clashing tones applied to objects which he will never get to relate to one another by reflections, and which remain crude, isolated, cold, and gaudy" (14). Ingres, defending his style, said: "The essential qualities of color are not to be found in the masses of lights and darks in the picture; they are rather in the brightness and individuality of the colors of objects" (15). Delacroix used optical mixture, complementary contrasts, and colored shadows, and believed in the importance of three primary colors. A recommendation to fellow painters was to "banish all earth colors" from the palette. His direct descendants were the impressionists, who created a revolution away from academic painting styles in battles that lasted more than 20 years.

An avid pupil (if implicit only) of Delacroix was the Dutch painter van Gogh. His palette underwent dramatic change during his brief lifetime, beginning with typically Dutch brownish, subdued colors. In Paris he came in contact with the impressionists and his palette began to lighten. Having access to Delacroix's notebooks, he began to accept some of the latter's rules, such as that good colorists (unlike Ingres) do not paint objects in local colors (their "true" colors). From Delacroix van Gogh learned the value of simultaneous contrast and how to use it for esthetic effect. But he concluded that he could understand Delacroix fully only by moving to the South of France, with its much different light atmosphere. Here he began to paint in pure paints as they came out of the tube. In one of his many letters to his brother Theo accompanying an order for paints, he explained his palette as consisting of three different chrome yellows (lemon, yellow, orange), Prussian blue, emerald green, Veronese green, madder red, and minium orange. Together with black and white these represented the paints for his works produced in the Provence. Softer or harder contrast effects were used by him to express the overall emotion suggested by the painting.

The beginnings of experimental psychology in the middle of the nineteenth century in Germany and the work of Maxwell and Helmholtz in color science helped sharpen the understanding by young artists of the fleeting nature of color. In 1879 the American physicist Rood published *Modern Chromatics*, a book describing the contemporary state of knowledge about color phenomena. It was soon translated into other languages, including French, and became an important source of scientific information about color, of interest to impressionist and postimpressionist painters. The name of the impressionist school of painting came from a work by Monet titled *Impression Sunrise*, exhibited in Paris in 1872 and causing an uproar among viewers and critics. It was described as vague and brutal and "worse than anyone so far has dared to paint." The changeable effects of varying natural lights on objects was demonstrated by Monet in his "Haystack" series and in the series depicting the cathedral in Rouen at different times of the day.

Impressionists generally applied color as they experienced it in nature and without theoretical concepts. Such concepts were of great interest to the postimpressionist Seurat. Each of his canvases is the result of a particular systematic set of ideas, both in design and in the application of color. He named his style of painting chromoluminarism; others called it pointillism. It consists (in his later works) of application of small dots of paint to the canvas (and even to the frame) in an attempt to create luminous intensity. He avidly studied the writings of Delacroix, Henry, Rood, and others as background for his own efforts. The application of dots of color can, when viewed at sufficient distance, produce additive color mixture, also employed in color television (but with lights). In the case of painted dots, the result is dulling and desaturation.

At the short-to-middle distance, the effect is one of scintillation and airiness. Seurat's friend Signac extended pointillism to divisionism where, instead of in the form of dots, paint was applied in the form of relatively large strongly colored strokes. The resulting optical effect is color interaction due to color contrast (16). The work of Signac and other divisionists led directly to that of the *fauves*, bringing us into the twentieth century. Color had already begun to move away from being naturalistic in the work of the divisionists and even more so in that of the *fauves*. At the same time, Cézanne also began to free form from naturalistic constraints, thereby opening the path for the abstract art of the twentieth century. Already in the middle of the nineteenth century photography assumed the role of the medium of naturalistic reproduction.

TWENTIETH CENTURY

With the beginning of the twentieth century and the continuing development of photography, the importance of representational art began to fade. In 1907 Picasso exhibited his important *Les Demoiselles d'Avignon*, and soon the style of cubism dissolved physical reality into abstraction.

Renewed interest in a theoretical foundation of the use of color in arts and crafts was generated at the Bauhaus in Germany, founded in 1919. Among its teachers were the artists Kandinsky, Klee, Schlemmer, Itten, and Albers. According to Kandinsky's essay *Über das Geistige in der Kunst* (On the spiritual in art) published in 1911, the purpose of art is to affect the viewer's soul with appropriate harmonies of color and form (17). Kandinsky posited three fundamental pairs of chromatic colors: yellow–blue, red–green, and orange–violet. Each pair has an antagonistic nature: for example, yellow is warm, moves toward the viewer, and is material and eccentric; its opposite blue is cold, moves away from the viewer, and is spiritual and concentric.

Another Bauhaus teacher, Paul Klee, expressed his views on form and color in a 1924 speech "On modern art," where he described the three important aspects of painting to be line, lightness–darkness, and color. His three fundamental pairs of chromatic colors (subtly different from Kandinsky's) are red–green, yellow–violet, and blue–orange, in line with recognized complimentary colors. He wrote of his art: "I have tried pure drawing, I have tried pure black-grey-white painting. In color I have experimented with all partial operations to which the study of the color circle have led me. I have elaborated color-weighted value painting, complementary color painting, colorful painting, and total-color painting" (18). Two other Bauhaus masters and practicing artists wrote programmatic and pedagogic essays on color: Itten's *The Art of Color*, and Albers's *Interaction of Color* (19). Both works remain in print.

In the twentieth century, color has become a completely free and independent creative tool for artists, at times not even restrained by form. Every known coloristic, optical, and psychological effect has been used by artists (see next section). The feeling of liberation experienced by the impressionists and their followers has dissipated. In the postmodern era, realism has made a comeback in many forms. Today, painting itself has lost much of its past position in favor of installations, videos, performance art, and other techniques, with no limits on subjects, materials, or tools.

OPTICAL AND PSYCHOLOGICAL EFFECTS IN PAINTING

Several of the fundamental optical and psychological effects also employed in painting have been described in Chapter 4. Other effects involve the implied movement of component forms, if highly chromatic colors of equal luminous reflectance are juxtaposed, or if there are strong lightness contrasts with repetitive patters, looseness of forms, and outlines to generate a feeling of action and movement, highlighting of certain aspects by local sharpening of contours and contrasts, perspective and illusion effects, color spreading effects from thin chromatic contours, forming of illusory contours as a result of strong lightness and chromatic contrasts, and others. These are sometimes consciously applied by the artists, but at other times seemingly intuitively (20).

It is evident that, while color is an essential component of painting and several crafts, theories of color and their strict or loose application have been of more than passing interest only to a relatively small number of artists. There is little question that while knowledge of the basic facts of color perception and colorant mixture can be very useful to an artist, it is not essential in any way for the creation of masterpieces. Art is usually created on an emotional level, with the artist working out the technical issues by trial and error. For painters it is useful to have knowledge about the following:

Facts of colorant mixture
Degree of opacity of each colorant used
Effect of transparent colorant when overlaid over opaque colorations
Effects of lightness and chromatic simultaneous contrast
Effect of the size and color of a colored field on the balance of the image
Effect of different light sources on the perceived color of the colorants in use
Color constancy and metamerism
Fastness and toxic properties of colorants.

12

Harmony of Colors

Harmonia, one of the reputed daughters of Aphrodite, goddess of beauty, lent her name to the esthetic principle of harmony. The *Oxford English Dictionary* defines harmony in the relevant context as "Combination of parts or details in accord with each other so as to produce an aestethically pleasing effect or agreeable aspect arising from apt arrangement of parts." The idea of harmony is very old, and harmony is a concept that has been closely associated with nature. There is a belief that everything that is truly natural is harmonious, in the universe as well as on Earth. In the West this was first conceptualized by Pythagoras (ca. 560–480 B.C.), the founder of a quasi-religious school. There are no texts extant that can be traced to him (Pythagoreans kept their knowledge secret), but Aristotle, in *Metaphysics*, reports:

> . . . the so-called Pythagoreans, who were the first to take up mathematics, not only advanced this study, but also having been brought up in it they thought its principles were the principles of all things. . . . they saw that the modifications and the ratios of the musical scales were expressible in numbers; . . . numbers seemed to be the first things in the whole of nature, they supposed the elements of numbers to be the elements of all things, and the whole heaven to be a musical scale and a number.

The Pythagoreans developed the original form of the *quadrivium*, the four mathematical sciences astronomy, arithmetic, geometry, and harmony. They believed numbers to be of divine origin and took certain numerical ratios as expressions of harmonic relationships. The fundamental image of the power of numbers is the *tetractys* (Fig. 12.1), the triangular arrangement representing the numbers 1 to 4 and at the same

Color: *An Introduction to Practice and Principles, Second Edition,* by Rolf G. Kuehni
ISBN 0471-66006-X Copyright © 2005 John Wiley & Sons, Inc.

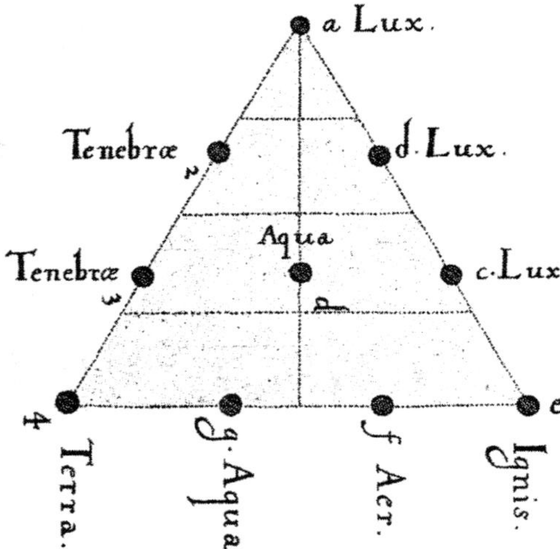

FIGURE 12.1 *The* tetractys *showing the development of the four classical elements from light (lux) and darkness (tenebrae) (1).*

time the sum 10. Pythagoras is said to have discovered the symphonic musical scale by attaching appropriate weights to the strings of a lyre. An octave was given the numerical ratio 2:1, and fifth and fourth were defined as 2:3 and 3:4, respectively. Such ratios were found in and applied to different Greek musical and poetry styles. The success in music of Pythagoras's idea was such that it influenced all subsequent writers on musical theory down to our time.

Aristotle was influenced considerably by the Pythagoreans. He discussed their musical theory and, in a cursory manner, also applied it to colors. In *Sense and Sensibilia* he wrote:

> Such then is a possible way of conceiving the existence of a plurality of colors besides the white and black; and we may suppose that many are the result of a ratio; for they may be juxtaposed in the ratio of 3 to 2 or 3 to 4, or in ratios expressible by other numbers; . . . and suppose that those involving numerical ratios, like the concords in music may be those generally regarded as most agreeable . . .

In the third century A.D. the neoplatonic philosopher Plotinus argued against Pythagorean ideas of harmony. Colors, according to him, had a beauty of their own that does not rely on harmonious relationships. Plotinus posited the One as the highest principle of the universe. An emanation from the One produces in the next realm intelligence that, in turn, emanates the soul. The One is represented by splendor, the highest level of light. Colors are lights, and the brighter they are, the more beautiful they are. In the fifteenth century the Italian neoplatonic philosopher Ficino established a qualitative Plotinian brightness scale of colors with white at the ninth and splendor at the twelfth and highest levels (see Chapter 10).

The two classical theories strongly influenced medieval thinking on beauty, and commentators such as Robert Grosseteste in the thirteenth century attempted to bring

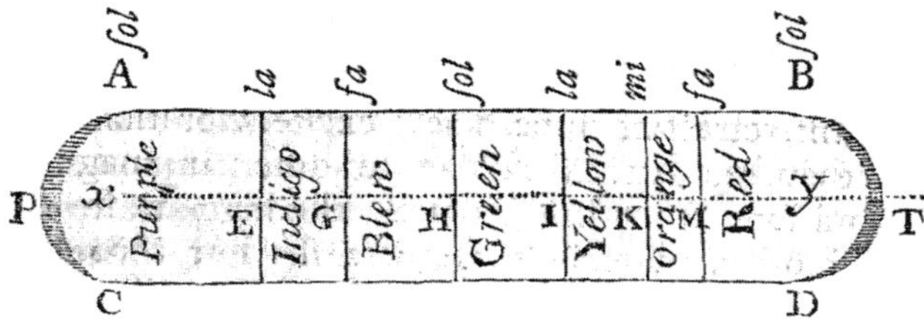

FIGURE 12.2 *Newton's spectrum colors divided according to the musical scale (1670s).*

them into agreement. The Pythagorean ideas on numerical harmony were transmitted into the Middle Ages by the late Roman philosopher Boethius (sixth century), who wrote important manuscripts on musical theory and geometry. As mentioned, he invented a graphical form of expressing musical harmony, a form that he also used to express other harmonic and logical relationships. This format was widely copied into the eighteenth century, and in the seventeenth also was used to express relationships between colors (see Fig. 10.2).

In the Renaissance there were many attempts to distill rules of color harmony from the writings of the ancients. However, as Gerard de Lairesse expressed it in *Het Groot Schilderboek* (1728): "It is strange that... so far nobody has provided a few basic rules [of color harmony] according to which one can proceed knowledgeably and with certainty." Many have tried to do so before and after de Lairesse, down to the recent past.

Despite best efforts, harmony, like beauty, has been found indefinable. Harmony tends to passivity and relaxation; it is an aspect of the contemplative life. Of taste the Romans said *de gustibus non est disputandum* (there is no arguing taste). The foundations of perceptions of beauty and harmony are unknown. Evolutionary theory, so far, has not provided any solid insights. There is no doubt that perceptions of beauty and harmony are strongly influenced by nurture and culture. History tells us that within a culture they change significantly over time. Illustrations of harmonious color combinations from 200, 100, or even 50 years ago are now often seen as trite, if not ugly. Depending on one's taste and upbringing, many convincing examples of the harmonious use of colors can be found. But unlike in physics and chemistry, where there are natural laws, there appear to be no natural laws of color or form harmony. The history of arts, crafts, and fashion illustrates the validity of this statement. Every culture and every school of art and design, or fashion, has attempted to define such rules, but these attempts lack generality.

Of the dozens of color systems proposed in the last 300 years, the majority has been developed with the idea that its particular form would be suitable to derive harmonic laws. Newton, a firm believer in universal harmony, selected his seven colors of the spectrum according to spacing, in agreement with the musical scale (Fig. 12.2). The systems of Runge, Chevreul, Brücke, Munsell, and Ostwald (and many more) were expressly developed to discover laws of harmony, but without any lasting effects. A modern Japanese collection of harmonious combinations (2) cannot be appreciated immediately by Western viewers because it is suffused with a Japanese esthetic. The growth of a world economy has begun to erase local and regional esthetics in favor of a more homogenized, international one. In today's international fashions, a discovery of new color harmonies developed in an isolated corner can soon become mainstream everywhere.

It is quite evident that there are no universal laws of harmony. Works of music and art, as history has taught us, do not depend on them. On the contrary, it can be argued that universal laws would be stifling, because limiting, for creativity. In the following no attempt is made to present a complete history of ideas on harmony, but rather to illustrate, in a few examples, some of the major ideas of the past.

COLOR AND MUSIC

Among the oldest ideas on color harmony is a presumed relationship between musical tones and colors. It is well known that some humans experience sounds, particularly those of letters, simultaneously as colors and vice versa. They are called *synesthetes*. This is an area of considerable current research in connection with the problem of consciousness (3). However, confirmed synesthetes experience individually different relationships between sounds and colors. The Pythagorean idea of universal harmony was celebrated in the book *Harmonices mundi* (World harmonies, 1619) by the astronomer Kepler, who also included Aristotle's seven-color scale among his examples. In 1740 the French Jesuit priest and mathematician Castel expanded Newton's seven spectral colors to eleven and worked on a color organ on which to play musical pieces as sequences of colors. The English pigment manufacturer George Field included in his book *Chromatics; or the Analogy, Harmony and Philosophy of Colours* (1817) an analogous scale of sounds and colors involving the enharmonic, chromatic, and diatonic scales. Detailed associations between sounds and colors had been proposed already in 1786 by the German painter Hoffmann, who also developed parallels between musical instruments and colors (4). In 1870 the German physicist Preyer described exactly eight different colors that can be seen in the spectrum, and defined these in terms of sounds, frequencies, and wavelengths (Fig. 12.3) (5). In 1854 the art historian Unger published a "chromharmonic" disk

Töne	Schwingungen		Farben	Wellenlänge in Milliont. Millim.	Fraunhofer's Linien, Wellenlänge n. Ångström.
	Intervalle	Billion in 1"			
c	1	388,2	braun	768,6	*A* 760,4
d	$\frac{9}{8}$	436,7	roth	683,2	*B* 686,7
e	$\frac{5}{4}$	485,2	orange	614,9	*C* 656,2
f	$\frac{4}{3}$	517,6	gelb	576,4	*D* 589,2
g	$\frac{3}{2}$	582,3	grün	512,4	*E* 526,9
a	$\frac{5}{3}$	647,0	blau	461,1	*F* 486,0
h	$\frac{15}{8}$	727,9	violett	409,9	*G* 430,7
c'	2	776,4	grau	384,3	H_2 393,3

FIGURE 12.3 *Preyer's table with tones* (Töne), *frequencies and interval ratios, colors* (Farben), *their wavelengths in nanometers, and the wavelengths of specific Fraunhofer lines in the spectrum (5).*

related to tones on which harmonious color combinations could be looked up. These consisted of complementary intervals as well as three- and four-color combinations (6). It is interesting to compare the widely divergent views of Hoffmann, Unger, and Preyer in the assignment of colors to tones:

Tone	Hoffmann	Unger	Preyer
c	dark blue	carmine	brown
d	violet	scarlet	red
e	red	yellow	orange
f	deep red	green	yellow
g	lemon yellow	ultramarine	green
a	grass green	violet	blue
h	sea green	brown red	violet

An ambitious theory of harmony was proposed in 1901 by the mineralogist Goldschmidt, who derived harmonic laws in sound and color from crystallography (7). Presumed relationships between music and color were also applied by abstract painters, such as Kandinsky and Mondrian. The *Prometheus Symphony* of the composer Scriabin included a color organ as an additional instrument when it was performed for the first time in New York in 1915. Color has remained an important aspect of the performance of pop music to this day; however, without any claims for direct association.

COMPLEMENTARY COLORS

Newton's incomplete color circle is already arranged so that compensative colors are approximately opposite, their mixture passing through white. In 1743 the French naturalist Comte de Buffon, author of a huge illustrated natural history, described detailed observations of simultaneous and successive contrasts in terms of colored shadows and aftereffects experienced after viewing colored fields. Color contrast effects in painting were already described by Leonardo, who named certain color pairs as contrary, for example, red–green, yellow–blue, and golden yellow–azure blue. Roger de Piles considered such combinations to be ugly. In the mid-eighteenth century, the painter Mengs based his color harmony theory on complementary colors. To aid painters Moses Harris placed in the color charts accompanying his book *The Natural System of Colours* of ca. 1770 "the most opposite, or contrary in hue" colors in opposing positions, thereby creating the model for the arrangement of most later color circles, for example, Runge's, whose complementary colors neutralize themselves (in theory) to neutral gray and who based his harmonic theory on complementaries.

A principle of harmonious colors based on complementaries was described in 1793 by Benjamin Thompson, the Count of Rumford. He proposed that colored lights are harmonious if together they combine to white. Even though Rumford only applied this rule to lights, it was soon also taken to apply in principle to colorants (8).

Harmony of complementary colors was also postulated by Goethe. Another rule of harmony, based on his six-color circle, involves pairs separated by an intermediate color. He described these as having character, but did not think them to be fully harmonious. According to Goethe, color combinations of adjacent pairs are "without character." Thus, there is a scale of declining harmony from perfection to character to without character. Lightness and darkness complicate harmonious relationships, but the mentioned rules are universally applicable. Active colors (yellow, orange, red) gain energy when combined with black or dark colors, but loose energy when combined with white or light colors. Passive colors (violet, blue, green) look dark and foreboding when combined with dark colors, but gain cheerfulness when combined with light colors.

Color contrast effects were investigated systematically by Chevreul, who described his results in 1839 in *De la Loi du Contraste Simultané des Couleurs* (On the law of simultaneous color contrast). Chevreul developed a more detailed theory of harmony involving contrast colors on the one hand and similar colors on the other. The former are not limited to opposing hues, but also can consist of strong lightness or saturation contrasts within the same hue. The latter involve nearby colors of the same hue, saturation, or lightness. Chevreul's work became very influential in the later nineteenth century.

COMPLEX RULES OF HARMONY

Among the authors of complex rules of color harmony, two are mentioned: Munsell and Ostwald. Munsell derived his balanced color sphere of 1906 for the express purpose of developing rules of harmony. He developed nine principles of harmony, among which are, for example, principle C: Opposite colors of equal chroma that center on middle value N5; or principle I: Harmony of the elliptical path (Fig. 12.4). He also included weight rules, such as: "Stronger chroma and value [colors] should

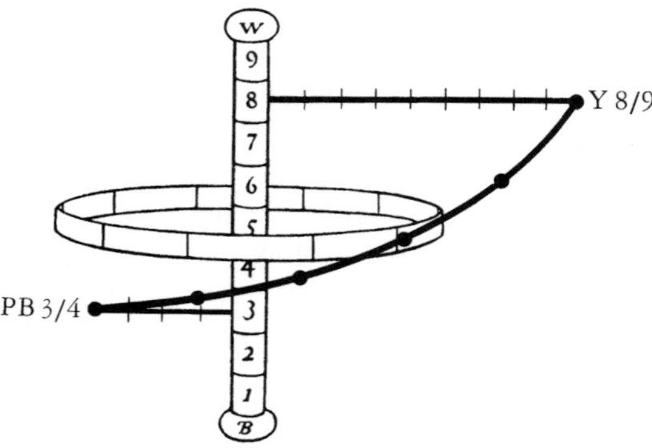

FIGURE 12.4 *Munsell's principle of harmony of the elliptic path in his balanced color sphere (9).*

occupy the lesser area and weaker chroma and lower value [colors] should occupy the greater area" (9).

In his book *Die Harmonie der Farben* (Harmony of colors, 1918) Ostwald described his key rule: "Colors appear to be harmonious or related if their properties are in certain simple relationships." Three different achromatic colors are required to obtain a harmonic relationship. The perceptual distances between the three must be identical. This applies also to four and five achromatic color harmonics. In the constant-hue triangles of the Ostwald system (see Chapter 5), he defines harmonic triples of constant whiteness, constant blackness, and constant purity by the rule applied to achromatic colors. For his color system, this results in 12,960 harmonic combinations. Harmonies combining achromatic and chromatic colors are also possible. This by no means exhausts all possibilities. Harmonies can be found within the 28 circles of constant whiteness and blackness in his double-cone color solid. Complementary colors represent a special case of these harmonies. Pairs with distances of three, four, five, six, eight, nine, or twelve colors on the 100-step hue circle are considered harmonious. An additional large number of harmonic combinations are obtained in the form of triplets and higher number combinations. In each situation Ostwald discusses the relationship of harmonic colors to tones. Ostwald and his pupils developed a number of tools for their rapid identification. His proposals did not have the expected general resonance, however. Already in 1912 the painter Kandinsky discounted any systematic objective attempts at color harmony by claiming "it is clear that harmony of color must rest only on the principle of touching the human soul" and that the guiding principle is "the principle of inner necessity . . . A painting is done when it is internally fully alive . . . colors must be used not because they exist in this chord in nature, or do not, but rather they are in this chord necessary in this painting" (10).

In his popular *Kunst der Farbe* (The art of color), Itten prefaced his comments on harmony with "Experience and tests concerning subjective color harmonies indicate that different persons can differ in their judgment of harmony or disharmony." According to Itten "harmony is equilibrium. Two or more colors are harmonious when their mixture results in a neutral gray" (Rumford's rule applied to paints). He noted that "all other color combinations are of an expressive or disharmonious nature. Very generally, it can be said that all complementary color pairs and all triples in relationship of an equilateral triangle, a square, or a rectangle in the twelve-color circle are harmonious" (11,12).

CREATE YOUR OWN HARMONIES

It is evident that rules of harmony vary widely and lack a solid basis. Opinions have changed often. Readers with an interest in color harmony should investigate the old theories or create their own harmonies. A useful tool is a color atlas or other systematic arrangement with many loose color chips. Readers with access to a computer with a good color monitor can easily study color arrangements for perceived harmony and develop their own rules.

Timetable for Color in Science and Art

Scientist, Philosopher	Color Theory	Art Theory	Artists and Works of Art
Pythagoras (ca. 560–ca. 480 B.C.)	Empedocles (ca. 495–ca. 435 B.C.)		
Plato (ca. 427–347 B.C.	Democritus (ca. 460–ca. 370 B.C.)		Polygnotus (ca. 500–ca. 440 B.C.)
	Aristotle (384–322 B.C.)		Apelles (fl. 330 B.C.)
	Theophrastus (ca. 372–ca. 287 B.C.)		
	Lucretius (ca. 99–ca. 55 B.C.), *De Rerum Natura*		Pompeii, Herculaneum (1st century B.C.)
	Pliny the Elder (ca. 23–79), *Historia Naturalis*		
Plotinus (205–270)			Church mosaics (400–700)
	Avicenna (980–1037)		Illuminated books
	Theophilus (ca. 1080–ca. 1125)		Cathedral windows (10th–11th centuries)
	Robert Grosseteste (ca. 1170–1253)		
Thomas Aquinas (1225–1274)	Roger Bacon (ca. 1216–1294)		Giovanni Cimabue (ca.1340–ca. 1302)
			Giotto (ca. 1266–ca. 1337)
William of Ockham (ca. 1280–ca. 1349)		Cennino Cennini (ca. 1370–1440), *Libro dell'Arte* (ca. 1400)	Jan van Eyck (ca. 1390–1441)
		Leon Battista Alberti (1404–1472), *Della Pittura* (1435)	Fra Angelico (ca. 1400–1455)
			Masaccio (1401–ca. 1428)
			Piero della Francesca (ca. 1420–1492)
			Sandro Botticelli (ca. 1444–1510)
		Leonardo da Vinci (1452–1519)	Mathias Grünewald (1475–1528)
		Michelangelo (1475–1564)	
Niccolò Macchiavelli (1469–1527)			Titian (ca. 1490–1576)
Nicolaus Copernicus (1473–1543)			
	Bernardino Telesio (1509–1588), *De Colorum Generatione* (1570)	Giorgio Vasari (1511–1574), *Vite . . .* (1550)	Franco–Flemish tapestries
Galileo Galilei (1564–1642)	Sigfrid Aronus Forsius (1560–1624), *De Colore* (1611)	Giovanni Paolo Lomazzo (1538–1600), *Trattato Dell'Arte* (1584)	
Johann Kepler (1571–1630)			Peter Paul Rubens (1577–1640)
	Franciscus Aguilonius (1567–1617), *Opticorum Libri Sex* (1613)	Raffaello Borghini (16th century)' *Il Riposo* (1584)	Frans Hals (ca. 1580–1666)
			Nicolas Poussin (1594–1665)

René Descartes (1596–1650)	Francis Glisson (ca. 1597–1677), *De Coloribus Pilorum* (1677)	Matteo Zaccolini (ca. 1574–1630), *Prospettiva del Colore*	Diego Velazquez (1599–1660) Rembrandt (1606–1669) Jan Vermeer (1632–1675)
Spinoza (1632–1677)	Athanasius Kircher (ca. 1601–1680), *Ars Magna Lucis et Umbrae* (1646) Isaac Vossius (1618–1689), *De Lucis Natura et Proprietate* (1662) Franciscus Maria Grimaldi (1618–1663), *Physico–Mathesis de Lumine, Coloribus et Iride* (1665) Robert Boyle (1627–1691), *Experimenta et Consideratione de Coloribus (1665)*	Roger de Piles (1635–1709), *Dialogue sur le Coloris* (1672) Henri Testelin (1616–1695), *Tables des Precept sur la Couleur* (1679)	
Gottfried von Leibniz (1646–1716)	Isaac Newton (1642–1727), *Opticks* (1704) Jakob Christof Le Blon (1667–1741), *Il Coloritto* (1725?) Louis Bertrand Castel (1688–1757), *L'Optique des Coùleurs* (1740)	Gerard de Lairesse (1641–1711), *Het Groot Schilderboek* (1707)	Giovanni Battista Tiepolo (1696–1770)
David Hume (1711–1776) Immanuel Kant (1724–1804)	Tobias Mayer (1723–1762), *De Affinitate Colorum Commentatio* (1758) Johann Heinrich Lambert (1728–1777) *Farbenpyramide* (1772)	Anton Raffael Mengs (1728–1779), *Gedanken über die Schönheit.* (1762)	Jean Baptiste Chardin (1699–1779)
Antoine Lavoisier (1743–1794)	Moses Harris (ca. 1731–ca. 1788), *The Natural System of Colours* (ca. 1770) George Palmer (18th century), *Theory of Colour and Vision* (1777) Johann Wolfgang Goethe (1749–1832), *Farbenlehre* (1810)		Francisco Goya (1746–1828) Jacques–Louis David (1748–1825)

(*Continued*)

Timetable for Color in Science and Art (*Continued*)

Scientist, Philosopher	Color Theory	Art Theory	Artists and Works of Art
Georg Wilhelm Hegel (1770–1831)	Thomas Young (1773–1829)		J. M. W. Turner (1775–1851)
	Phillip Otto Runge (1777–1810), *Farben–Kugel*		John Constable (1776–1837)
Artur Schopenhauer (1788–1860)	Michel-Eugène Chevreul (1786–1889), *De la Loi du Contraste Simultané des Couleurs* (1839)		Eugene Delacroix (1798–1863)
Charles Darwin (1809–1882)	Hermann Günter Grassmann (1809–1877)		
	Hermann von Helmholtz (1821–1894), *Handbuch der Physiologischen Optik* (1867)		
Dimitri Mendeleev (1834–1907)	James Clerk Maxwell (1831–1879)	Ogden Nicholas Rood (1831–1902), *Modern Chromatics* (1879)	Édouard Manet (1832–1883)
	Ewald Hering (1834–1918), *Grundzüge der Lehre vom Lichtsinn* (1920)		Paul Cézanne (1839–1906)
			Paul Gaugin (1848–1903)
Friedrich Nietzsche (1844–1900)	Johannes von Kries (1853–1928)		Vincent van Gogh (1853–1890)
	Albert Munsell (1858–1918), *A Color Notation* (1905)		Georges Seurat (1859–1891)
		Wassily Kandinsky (1866–1944), *Über das Geistige in der Kunst* (1912)	
	Wilhelm Ostwald (1853–1932), *Farbenlehre* (1918–1922)		Emil Nolde (1867–1956)
			Henri Matisse (1869–1954)
Bertrand Russell (1872–1970)		Paul Klee (1879–1940), *Über die Moderne Kunst* (1924)	
Albert Einstein (1879–1955)			
Ludwig Wittgenstein (1889–1951)	Erwin Schrödinger (1887–1961)	Josef Albers (1888–1976), *Interaction of Color* (1963)	Piet Mondrian (1872–1944)
	Georg Elias Müller (1850–1934), *Uber die Farbenempfindungen* (1930)		Pablo Picasso (1881–1973)
Niels Bohr (1885–1962)		Johannes Itten (1889–1967), *The Art of Color* (1961)	George Braque (1882–1963)
	Dean Brewster Judd (1900–1972)		Jackson Pollock (1912–1956)
Kurt Gödel (1906–1978)	William David Wright (1906–1997)		Mark Rothko (1903–1970)
Jonas Salk (1914–1995)	Manfred Richter (1905–1990)		Willem de Kooning (1904–1997)
	David Lewis MacAdam (1910–1998)		

Notes

Chapter 1

1. The definition is from *Merriam-Webster's Collegiate Dictionary*, 10th ed., Def. 1a, 2001.
2. *International Lighting Vocabulary*, 3rd ed., Publication CIE No. 17, Vienna: CIE, 1970.
3. K. Nassau, *The Physics and Chemistry of Color*, New York: Wiley, 1983. See also The physics and chemistry of color: the 15 mechanisms, in The Science of Color, 2nd ed., S.K. Shevell, ed., Amsterdam: Elsevier, 2003.
4. The Kelvin scale is the absolute temperature scale named after the British physicist W. T. Kelvin (1824–1907), abbreviated K, beginning at the lowest possible temperature: $0\ K = -273.15°C$. The units are of the same size as those of the Celsius scale.
5. The word "Reflection" is from the Latin *reflectere*, meaning to break. In the Middle Ages the science of reflection was known as cathoptrics.
6. Lambert determined in the eighteenth century that equal total thickness of a layer, regardless of the number of layers, causes equal light absorption. Beer's law states that equal amounts of dye in the solution cause equal absorption. Together, they form what is known as the Lambert–Beer law; see also Chapter 8.
7. Refraction, from the Latin *refrangere*, meaning to bend back. The science of refraction was known as dioptrics in the Middle Ages.
8. Conjugate, from the Latin *conjugare*, meaning to yoke together. An organic substance containing conjugated bonds has two or more double bonds separated by single bonds.
9. *Colour Index*, Fourth Edition Online, Society of Dyers and Colourists and American Association of Textile Chemists and Colorists, 2002. The *Index* is a compilation of colorants available on the world market. Entities are designated by a Color Index name, such as C. I. Disperse Red 60.

Color: *An Introduction to Practice and Principles, Second Edition*, by Rolf G. Kuehni
ISBN 0471-66006-X Copyright © 2005 John Wiley & Sons, Inc.

10. Nassau, *Physics and Chemistry of Color.*
11. Laser is an acronym for light amplification by stimulated emission of radiation.

Chapter 2

1. An up-to-date book on the general subject of vision is S. E. Palmer, *Vision Science*, Cambridge: MIT Press, 2002. The current state of knowledge about the neurobiology of vision is described in C. Koch, *The Quest for Consciousness,* Englewood, CO: Roberts, 2004.
2. For an interesting discussion, see T. Nørretranders, *The User Illusion*, New York: Viking, 1998.
3. For impressive examples of change blindness, visit Web sites www.personal.kent.edu/%7Edlevin/!homepage.htm; http://nivea.psycho.univ-paris5.fr/; and http://viscog.beckman.unic.edu/djs_lab/demos.html.
4. Recent papers on the evolution of color vision are: J. Nathans, The evolution and physiology of human color vision: Insights from molecular genetic studies of visual pigments, *Neuron* 24 (1999) 299–312; B. Wissinger and L. T. Sharpe, New aspects of an old theme: The genetic basis of human color vision, *American Journal of Human Genetics* 63 (1998) 1257–1262; A. K. Surridge, D. Osorio, and N. I. Mundy, Evolution and selection of trichromatic vision in primates, *Trends in Ecology and Evolution* 18 (2003) 198–205.
5. For supporting data, see, for example: B. C. Regan, C. Juillot, B. Simmen, F. Viénot, P. Charles-Dominique, and J. D. Mollon, Fruits, foliage and the evolution of primate colour vision, *Philosophical Transactions of the Royal Society London* B 356 (2001) 229–283.
6. For different eye types, see Web site: www.ebiomedia.com and look up the essay "Eye to Eye" in "Galleries," or site www.maayan.uk.com/evoeyes2.html.
7. M. Neitz, T. W. Kraft, and J. Neitz, Expression of L-cone pigment gene subtypes in females. *Vision Research* 28 (1998), 3221–3225.
8. See, for example, P. G. Kevan and W. G. K. Backhaus, Color vision: Ecology and evolution in making the best of the photic environment. *In* W. G. K. Backhaus, R. Kliegl, and J. S. Werner, eds., *Color Vision*, Berlin: Walter de Gruyter, 1998, and F. G. Barth, *Insects and Flowers*, Princeton, NJ: Princeton University Press, 1985.
9. B. Russell, *The Problems of Philosophy*, London: Oxford University Press, 1912.
10. For an interesting discussion of this and connected issues, see M. Ridley, *Nature via Nurture*, New York: Harper Collins, 2003.
11. Recent synesthesia literature: R. Cytowic, *Synesthesia: A Union of the Senses*, 2nd ed., Cambridge, MA: MIT Press, 2002; S. Baron-Cohen, ed., *Synesthesia: Classic and Contemporary Readings*, Oxford: Blackwell, 1997; J. Harrison, *Synaesthesia: The Strangest Thing*, Oxford: Oxford University Press, 2001.
12. For an excellent introduction to the issues of consciousness, see S. Blackmore, *Consciousness*, Oxford: Oxford University Press, 2003.
13. For an overview, see B. Maund's article Color, in the Internet Stanford University Encyclopedia of Philosophy at http://plato.stanford.edu/entries/color/.
14. An up-to-date exchange of opinions on this subject is found in A. Byrne and D. R. Hilbert, Color realism and color science, *Behavioural and Brain Sciences* 26 (2003) 3–63.
15. D. L. MacAdam, Note on the number of distinct chromaticities, *Journal of the Optical Society of America* 37 (1947) 308–309.

Chapter 3

1. More details on the retina are found, for example, in R. H. Masland, The fundamental plan of the retina, *Nature Neuroscience* 4 (2001) 877–886.
2. For highly technical details, see several chapters in K. R. Gegenfurtner and L. T. Sharpe, eds., *Color Vision*, Cambridge: Cambridge University Press, 1999.
3. R. A. Rensink, Seeing, sensing, and scrutinizing, *Vision Research* 40 (2000) 1469–1487.
4. On rhodopsin, consult, for example, Web site www.chemsoc.org/exemplarchem/entries/2002/upton/rhodopsin.htm.
5. From D. Jameson, *Handbook of Sensory Physiology*, Vol. 7/4, Berlin: Springer-Verlag, 1972.
6. For more details on rods and cones, see, for example, Web site http://webvision.med.utah.edu/photo2.html.
7. More details on the physiology of color vision are found in the book listed in Note 2. For additional, more detailed information on the anatomy and physiology of the visual system, see, for example, Web site www.webvision.med.utah.edu.
8. On hue cancellation experiments, see L. M. Hurvich, *Color Vision*, Sunderland MA, Sinauer, 1981.

Chapter 4

1. On light versus surface perception see, for example, R. Mausfeld, "Colour" as part of the format of different perceptual primitives: the dual coding of colour. In R. Mausfeld and D. Heyer, eds., *Colour Perception*, Oxford: Oxford University Press, 2003.
2. Room foreshortening: reported in S. Krauss, Tatsachen und Probleme zu einer psychologischen Bedeutungslehre auf Grundlage der Phänomenologie, *Archiv für die gesamte Psychologie* 62 (1928) 179–225.
3. D. Purves and R. B. Lotto, *Why We See What We Do*, Sunderland MA: Sinauer, 2003.
4. For results of an analysis of HKE experimental data, see R. G. Kuehni, *Color Space and Its Divisions*, Hoboken, NJ: Wiley, 2003.
5. An early investigation of lightness crispening is that by T. Kaneko, A reconsideration of the Judd-Cobb lightness function, *Acta Chromatica* 1 (1964) 103–110. The effect of the surround lightness on perceived lightness has been described in a graph based on Munsell-value samples in D. B. Judd and G. Wyszecki, *Color in Business, Science, and Industry*, 3rd ed., New York: Wiley, 1975, p. 291.
6. For a compilation of experimental data, see R. G. Kuehni, Variability in unique hue selection: a surprising phenomenon, *Color Research and Application*, 29 (2004) 158–162.
7. *ASTM E284. Standard Terminology of Appearance*, West Conshohocken, PA: American Society of Testing Materials, 1996.
8. See reference in Note 3.
9. For a description of zero grayness results, see R. M. Evans, *The Perception of Color*, New York: Wiley, 1974.
10. The desaturation of color is conceptually described in P. O. Runge, *Die Farben-Kugel*, Hamburg: Perthes, 1810.
11. Several color appearance models are described and compared in M. D. Fairchild, *Color Appearance Models*, 2nd ed., Reading, MA: Addison Wesley Longman, 2004.

12. The color inconstancy index is described in M. R. Luo and R. W. Hunt, A chromatic adaptation transform and a colour inconstancy index, *Color Research and Application* 23 (1998) 154–158.
13. The color rendering index is described, for example, in G. Wyszecki and W. S. Stiles, *Color Science*, 2nd ed., New York: Wiley, 1982.
14. For examples of simultaneous contrast, visit www.uni-mannheim.de/fakul/psycho/irtel/cvd. html.
15. For an English version of Chevreul's book, see M. E. Chevreul, *The Principles of Harmony and Contrast of Colors and their Application to the Arts*, West Chester; PA: Schiffer, 1987. Chevreul was interested in finding the reason for the considerable changes in apparent hue and lightness of dyed yarns when woven into tapestries.
16. For an impressive example of colored shadow, visit the Web site www.kyb.tuebingen.mpg.de/~wehrhahn and click on "colored shadow."
17. For successive contrast demonstrations, visit http://wisebytes.net/illusions/afterimages.php.
18. For information on the Saybolt Chromometer, visit koehlerinstrument.com. The platinum–cobalt scale is described in ASTM test method D1209.97 *Standard Test Method for Color of Clear Liquids*.

Chapter 5

1. This result was obtained from 37 trichromatic females and males as described in K. A. Jameson, S. M. Highnote, and L. M. Wasserman, Richer color experience in observers with multiple photopigment opsin genes, *Psychonomic Bulletin and Review* 8 (2001) 244–261.
2. Isaac Newton, *Opticks*, London: Smith and Walford, 1704.
3. Quantitative analysis of the chromatic power of pigments was initiated in the late nineteenth century by Abney and others.
4. The Nickerson formula is found in D. Nickerson, The specification of color tolerances, *Textile Research* 6 (1936) 509–514.
5. A detailed description of hue superimportance is found in D. B. Judd, Ideal color space, *Palette* 29 (1969) 25–31; 30: 21–28; 31: 23–29.
6. For a brief, modern description of scaling, see L. E. Marks and D. Algom, Psychophysical scaling. In M. H. Birnbaum, ed., *Measurement, Judgment, and Decision Making*, New York: Academic Press, 1998.
7. References to the fact that OSA-UCS has been modified to fit into a Euclidean space are scarce in the literature. The most significant is found in D. Nickerson, History of the OSA Committee on Uniform Color Scales, *Optics News*, Winter 1977, 8–17.
8. For a more extended discussion of this and other subjects related to color spaces and scaling, see R. G. Kuehni, *Color Space and Its Divisions*, Hoboken, NJ: Wiley, 2003.
9. W. Schönfelder, Der Einfluss des Umfeldes auf die Sicherheit der Einstellung von Farbgleichungen, *Zeitschrift für Sinnesphysologie* 63 (1933) 228–236.
10. See the reference in Note 8 for more details on the factor 4 in size of unit color difference ellipses as a function of chroma.
11. The history of the development of NCS was described by A. Hård, G. Tonnquist, and

L. Sivik, NCS, Natural Color System – From concept to research and applications. Parts I and II, *Color Research and Application* 21 (1996) 180–205 and 206–220.

12. The history of the Munsell system has been described by: D. Nickerson, History of the Munsell system, company, and foundation, *Color Research and Application* 1 (1976) 7–10; R. S. Berns and F. W. Billmeyer, Development of the 1929 Munsell Book of Color: a historical review, *Color Research and Application* 10 (1985) 246–250; R. G. Kuehni, The early development of the Munsell system, *Color Research and Application* 27 (2002) 20–27.
13. Definitions of the aim colors of the Munsell Renotations can be found in G. Wyszecki and W. S. Stiles, *Color Science*, 2nd ed., New York: Wiley, 1982.
14. C. E. Foss, Space lattice used to sample the color space of the Committee on Uniform Color Scales of the Optical Society of America, *Journal of the Optical Society of America* 68 (1978) 1616–1619.
15. OSA-UCS was described by D. L. MacAdam, Uniform color scales, *Journal of the Optical Society of America* 64 (1974) 1691–1702. The aim-color definitions are found in Wyszecki and Stiles, see Note 12. The American Society of Testing Materials has published a *Standard Practice for Specifying Color by Using the Optical Society of America Uniform Color Scales System*, West Conshohocken, PA: American Society of Testing Materials, 1996.
16. The history of OSA-UCS was described by D. Nickerson in the article referenced in Note 7.
17. The figure has been reproduced from D. Nickerson, OSA Uniform Color Scale Samples: A unique set, *Color Research and Application* 6 (1981) 1–33.
18. Ridgway published a small early set of color samples in 1886 and the larger system as R. Ridgway, *Color Standards and Color Nomenclature*, Washington, DC: published by the author, 1912.
19. The DIN 6164 system was developed by M. Richter and an atlas first issued in 1984. Its development has been described in M. Richter and K. Witt, The story of the DIN color system, *Color Research and Application* 11 (1986) 138–145.
20. The *Eurocolor* system was published in 1984, but is no longer available. The *ACC* system was developed by the Dutch paint manufacturer Sikkens in 1978. *Colorcurve* was first published in 1988 by Colorcurve Systems, Inc., in Minneapolis, MN.
21. US Patent 3,474,546 for Visual Arts Matching Charts was issued in 1969 to C. J. Wedlake. H. Küppers, *DuMont's Farben-Atlas*, Köln: DuMont, 1978.
22. Dimension reduction of spectral power and reflectance functions have been an active field of study since the 1960s. A brief review of methods and results is found in R. Ramanath, R. G. Kuehni, W. E. Snyder, and D. Hinks, Spectral spaces and color spaces, *Color Research and Application*, 29 (2004) 29–37.
23. J. B. Cohen, Dependency of the reflectance curves of the Munsell color chips, *Psychonomic Science* 1 (1964) 369–370.
24. The figure is copied from R. Lenz, M. Österberg, J. Hiltunen, T. Jaaskelainen, and J. Parkkinen, Unsupervised filtering of color spectra. *Journal of the Optical Society of America A* 13 (1996) 1315–1324.
25. The theory was developed by Berlin and Kay: B. Berlin and P. Kay, *Basic Color Terms*, Berkley CA: University of California Press, 1969.
26. NBS Circular 553 *The ISCC-NBS Method of Designating Colors and a Dictionary of Color Names*, Washington, DC: National Bureau of Standards, 1955.

Chapter 6

1. Spectral power distributions are numbers or their plots representing the power of a light source at a given wavelength relative to its power at 555 nm = 100. The unit of radiant power or energy is the joule or the watt. Radiant intensity is expressed in watts per unit spherical space segment (solid angle).
2. Illuminant F12 represents a so-called triband fluorescent light, so named because its energy is radiated primarily in three narrow bands. This and the fact that little or no energy is radiated outside the visible spectrum makes these lamps comparatively highly energy efficient.
3. The Scottish physicist J. C. Maxwell (1831–1879) conducted color-matching experiments using a mixture of color stimuli from rapidly rotating disks (so-called Maxwell disks) to determine the color fundamentals as well as, later, a visual colorimeter. Important early determinations using a visual colorimeter were made by Helmholtz's assistant A. König and his collaborator C. Dieterici in 1888.
4. J. Guild, The colorimetric properties of the spectrum, *Philosophical Transactions of the Royal Society (London)* A230 (1931) 149–187.
5. There are four laws of Grassmann. The symmetry law states that if stimulus **A** matches stimulus **B**, then **B** must also match **A**. According to the transitivity law, if **A** matches **B** and **B** matches **C**, then **A** also matches **C**. The proportionality law indicates that if **A** matches **B**, then $a\mathbf{A}$ matches $a\mathbf{B}$, where a is any positive factor adjusting the radiant power of the lights up or down. The additivity law states, for example, that if **A** matches **B** and **C** matches **D**, then $(\mathbf{A} + \mathbf{D})$ matches $(\mathbf{B} + \mathbf{C})$. The first three laws have been experimentally found to generally apply, while the fourth law only applies under certain conditions.
6. In recent years this issue has come under discussion again, and a CIE research committee has been formed to experimentally address it again. Given that there are biological processes involved, it is unlikely that Grassmann's laws are followed with engineering precision. As a result, in extreme testing conditions, it may be found that Grassmann's laws are followed less than perfectly.
7. CIE No. 15.2 *Colorimetry*, 2nd ed., Vienna: CIE, 1986. The system is also extensively described in G. Wyszecki and W. S. Stiles, *Color Science*, 2nd ed., New York: Wiley, 1982.
8. G. Wyszecki and W. S. Stiles, *Color Science*, 2nd ed., New York: Wiley, 1982.
9. Note that the author here conveniently and easily, but with only loose justification, began to combine appearance terminology with stimulus definition terminology. This is indicative of the historical process that has taken place and that resulted in the mushy language most everybody is using today in this field.
10. There are several possible reasons for such changes that have to do with a layer of yellow absorbing pigment (macula) surrounding the fovea in the retina or with the possible angles at which photons can interact with cones.
11. In the author's opinion, it is regrettable that the CIE, presumably for reasons of geometrical convention, did not select the x and z scales as coordinates of the chromaticity diagram. They are purer representatives of chromatic information.
12. The chemist W. Ostwald (1853–1932) wrote a multivolume book on color science and developed a color order system named after him. The physicist E. Schrödinger (1887–1961) wrote two important papers on color stimuli and their relationship to perceived color in 1920 and 1926.

13. Such solids had been calculated by Luther and Rösch in Germany and by Nyberg in Russia before the CIE colorimetric system was established.
14. J. Koenderink and A. Kappers, *Color space*, Report 16/96 Center for Interdisciplinary Research, Bielefeld: University of Bielefeld, 1996.

Chapter 7

1. E. Q Adams, X-Z planes in the 1931 I.C.I. system of colorimetry, *Journal of the Optical Society of America* 32 (1942) 168–173.
2. S. S. Stevens, *Psychophysics*, New York: Wiley, 1975.
3. L. G. Glasser, A. H. McKinney, C. D. Reilly, and P. D. Schnelle, Cube-root color coordinate system, *Journal of the Optical Society of America* 48 (1958) 736–740.
4. CIE 1976 (L*a*b*)-System, in *Colorimetry*, 2nd ed., Vienna: CIE, 1986. Also discussed in G. Wyszecki and W. S. Stiles, *Color Science*, 2nd ed., New York: Wiley, 1982.
5. D. B. Judd, Chromatic sensibility to stimulus differences, *Journal of the Optical Society of America* 22 (1932) 72–108.
6. D. L. MacAdam, Uniform color scales, *Journal of the Optical Society of America* 64 (1974) 1691–1702.
7. H. v. Helmholtz, *Handbuch der physiologischen Optik*, 2nd ed., Hamburg: Voss, 1896.
8. D. L. MacAdam, Visual sensitivities to color differences in daylight, *Journal of the Optical Society of America* 32 (1942) 247–274.
9. For discussion of this and many other related issues, see R. G. Kuehni, *Color Space and Its Divisions*, Hoboken, NJ: Wiley, 2003.
10. G. Wyszecki and W. S. Stiles, *Color Science*, 2nd ed., New York: Wiley, 1982.
11. M. R. Luo, G. Cui, and B. Rigg, The development of the CIE 2000 colour-difference formula: CIEDE2000, *Color Research and Application* 26 (2001) 340–350.
12. Up to now lightness crispening has not been considered explicity in color difference formulas.
13. For a possible explanation of the tilt of ellipses of blue color stimuli, see R. G. Kuehni, Towards an improved uniform color space, *Color Research and Application* 24 (1999) 253–265.

Chapter 8

1. The Lambert–Beer law is named after the eighteenth century Swiss physicist J. H. Lambert (1728–1777) and the German physicist A. Beer (1825–1863), the former contributing the relationship between thickness of layer and absorption, the latter that of concentration and absorption.
2. Of the natural dyes mentioned indigo, madder, gamboge, and saffron are of plant origin, purple is from shellfish, and kermes and cochineal from insects.
3. For the story of Perkin, see S. Garfield, *Mauve*, New York: Norton, 2000.
4. D. L. MacAdam, *Color Measurement*, Berlin: Springer-Verlag, 1981.
5. According to Greek legend, minium was discovered when lead white in a pot in the house

of a painter that burned down was found to have turned bright red. Minium is believed to have given the name to miniature, a kind of painting used in medieval manuscripts.

6. Kubelka and Munk were German physicists who developed the so-called two-flux theory of absorption and scattering: P. Kubelka and F. Munk, Ein Beitrag zur Optik der Farbanstriche, *Zeitschrift für technische Physik* 12 (1931) 593–601. The theory is discussed in some detail in D. B. Judd and G. Wyszecki, *Color in Business, Science, and Industry*, 3rd ed., New York: Wiley, 1975.
7. More complex scattering models are four-flux, or multiflux, models, and the Mie theory. For further discussion, see, for example, J. C. Guthrie and J. Moir, The application of color measurement, *Review of Progress in Coloration* 9 (1978) 1–12.
8. Determination of K and S values of pigments is described, for example, in J. H. Nobbs, Colour-match prediction for pigmented materials. In R. McDonald, ed., *Colour Physics for Industry*, 2nd ed., Bradford, England: Society of Dyers and Colourists, 1997.
9. R. M. Johnston Color theory. In T. A. Lewis, ed., *Pigments Handbook*, Vol. III, New York: Wiley, 1973.
10. The terms tint, shade, and tone have had various meanings in the history of coloration. In this text the following definitions are used: tint (series), a series of colors where the full color (color at maximum chroma) has been diluted with increasing amounts of white; shade (series), a series of colors where the full color has been diluted with increasing amounts of black; tone (series), a series where a full, tint, or shade color has been diluted with increasing amounts of gray of the same lightness.
11. F. W. Billmeyer and M. Saltzman, *Principles of Color Technology*, 2nd ed., New York: Wiley, 1981.
12. Examples of fluorescent minerals can be found, for example, at the Web sites *http://home.pacifier.com/ ~leopard/* or *http://users.rcn.com/kenx/*.

Chapter 9

1. A technical text on color television is R. L. Hartwig, *Basic TV Technology: Digital and Analog*, 3rd ed., Focal Press, 2000.
2. For details on color photography technology, see R. W. G. Hunt, *The Reproduction of Colour in Photography, Printing, and Television*, 5th ed., England: Fountain Press, 1995.
3. For a useful text on digital photography, see J. A. King, *Digital Photography for Dummies*, 4th ed., Hoboken, NJ: Wiley, 2002.
4. For information on halftone and graphic printing in general, see: K. Johannsson, *A Guide to Graphic Printing Production*, Hoboken, NJ: Wiley, 2003.
5. R. S. Berns, *Billmeyer and Saltzman's Principles of Color Technology*, 3rd ed., Hoboken, NJ: Wiley, 2000.
6. An up-to-date text on color management is, A. Sharma, *Understanding Color Management*, Independence, MO: Delmar, 2003.
7. For more information on computer colorant formulation, see R. McDonald, ed., *Colour Physics for Industry*, 2nd ed., Bradford, England: Society of Dyers and Colourists, 1997, and R. G. Kuehni, *Computer Colorant Formulation*, Lexington, MA: Lexington Books, 1975.
8. For information on appearance measurement, see R. S. Hunter and R. W. Harold, *The Measurement of Appearance*, 2nd ed., New York: Wiley, 1987.

Chapter 10

1. Information on the Chauvet cave can be found in J. M. Chauvet, E. B. Deschamps, and C. Hillaire, *Dawn of Art: The Chauvet Cave*. New York: Abrams, 1996.
2. There are many books on ancient Egyptian art, for example, G. Robins, *The Art of Ancient Egypt*, Cambridge, MA: Harvard University Press, 1997.
3. Plutarch quote from J. Mansfeld, *Die Vorsokratiker I and II*, Stuttgart: Reclam, 1986. English translation by the author.
4. Democritus quote from K. Freeman, *Ancilla to the Pre-Socratic Philosophers*, Cambridge, MA: Harvard University Press, 1957.
5. Plato mixture table derived from statements in *Timaeus* 67 to 68. *In* Plato, *The Collected Dialogues*, E. Hamilton and H. Cairns, eds., Princeton, NJ: Princeton University Press, 1961.
6. An interesting source for information on the use of color terms by ancient Greek authors is E. Veckenstedt, *Geschichte der Griechischen Farbenlehre*, Paderborn, Germany: Schöningh, 1888.
7. Batman's translation is available in *Batman uppon Bartholome His Booke De Proprietatibus Rerum* 1582. Hildesheim, Germany: Olms, 1976.
8. L. Dolce, *Dialogo di M. Lodovico Dolce nel quale si ragiona delle qalità, diversità e proprietà dei colori* 1565. Accessible at www.bivionline.it.
9. For an extended discussion on Bacon's work on color, see C. Parkhurst, Roger Bacon on color. *In* K. L. Selig and S. Sears, eds., *The Verbal and the Visual*, New York: Italica Press, 1990.
10. Ficino and Cardano: M. Ficino, *Opera*, Vol. I, Basel 1519; H. Cardanus, *Hieronimi Cardani Opera Omnia*, Vol. 2, Lyon, 1563.
11. Theophilus's work is translated in Theophilus, *The Various Arts*, C. R. Dodwell, ed. and transl., Oxford: Clarendon Press, 1986.
12. Kircher was a man of wide knowledge. Born in Germany, he was later called to Rome to teach. His work on color is A. Kircher, *Ars Magna Lucis et Umbrae*, Amsterdam: Jansson and Waesberge, 1671. The work of d'Aguilon is F. Aguilonius, *Opticorum Libri Sex*, Antwerp: Plantin, 1613 (with illustrations by the painter P. P. Rubens). D'Aguilon and Kircher appear to have taken the idea of five simple colors from V. A. Scarmilionius, *De Coloribus*. Marburg, Germany, 1601.
13. On Forsius, see C. Parkhurst and R. L. Feller, Who invented the color wheel? *Color Research and Application* 7 (1982) 217–230.
14. The original text of Glisson is found in F. Glisson, *Tractatus de ventriculo et intestinis*, London: Brome, 1677. For an interpretation, see R. G. Kuehni and R. Stanziola, Francis Glisson's color specification system of 1677, *Color Research and Application* 27 (2002) 15–19.
15. F. M. Grimaldi, *Physico-Mathesis de lumine, coloribus, et iride*, Bologna: Victor Benati, 1665.
16. A Letter of Mr. Isaac Newton, Professor of the Mathematicks in the University of Cambridge; containing his New Theory about Light and Colors, *Philosophical Transactions of the Royal Society London*, February 19, 1671/2 (available on www.newtonproject.ic.ac.uk).
17. Le Blon published the results of his invention in J. C. le Blon, *Coloritto, or the Harmony of Colouring in Painting Reduced to mechanical Practice*, London [1725]. An image of

prints of three plates individually and combined is found in J. Gage, *Color and Culture*, Boston: Little, Brown, 1993.

18. Palmer was an English glassmaker with good connections to the French court. His book *Theory of Colours and Vision* was translated into French. In 1786 he published *Théorie de la Lumiere Applicable aux Arts* in Paris. This book contains a color mixture theory and a plan to publish a color atlas with 2160 samples. No copy of the atlas is known. He also offered lamps with blue glass filters that were to make lamplight look like daylight.
19. The color circles only appeared in the 1708 edition of *Traité de la Peinture en Mignature* published by van Dole in The Hague in 1708 and a later translation of the work into Dutch.
20. There are only four copies of Harris's book extant, one of which is in the Birren Library at Yale University. The book is believed to have been published ca. 1770. A facsimile (but with changes in the color plates) was privately printed by Faber Birren in 1963.
21. Mayer's essay on color order was published in *Opera inedita Tobiae Mayeri* in Göttingen in 1775. An English translation is available as A. Fiorentini and B. B. Lee, Tobias Mayer's On the relationship between colors, *Color Research and Application* 25 (2000) 66–74.
22. J. H. Lambert, *Beschreibung einer mit dem Calauischen Wachse ausgemalten Farbenpyramide*, Berlin: Haude und Spener, 1772.
23. P. O. Runge, *Die Farben-Kugel oder Construction des Verhältnisses aller Mischungen der Farben zueinander*, Hamburg: Perthes, 1810.
24. On Doppler and Schrödinger, C. Doppler, Versuch einer systematischen Classification der Farben, *Abhandlungen der königlichen bömischen Gesellschaft der Wissenschaften* 5 (1848) 401–412; E. Schrödinger, Grundlinien einer Theorie der Farbmetrik im Tagessehen, *Annalen der Physik* 63 (1920) 297–447 and 448–520.
25. For more detailed information, see R. G. Kuehni, *Color Space and Its Divisions*, New York: Wiley, 2003.
26. Figure from J. C. Maxwell, On the theory of compound colours, and the relations of the colours of the spectrum, *Proceedings of the Royal Society London* 10 (1860) 57–84.
27. The first to explicitly refer to three attributes in relation to lights was H. G. Grassmann. His proposals were modified slightly by Helmholtz in his *Physiologische Optik* in 1860.
28. E. Hering's final work, *Grundzüge der Lehre vom Lichtsinne*, was published between 1905 and 1911 by Springer in Berlin. It was translated by L. M. Hurvich and D. Jameson as *Outlines of a Theory of the Light Sense*, Cambridge, MA: Harvard University Press, 1964.
29. For a detailed account of the battles between Helmholtz and Hering, read R. S. Turner, *In the Eye's Mind*, Princeton, NJ: Princeton University Press, 1994.
30. References to Chevreul, Benson, and Kirschmann are as follows: M. E. Chevreul, *De la Loi du Contraste Simultané des Couleurs*, Paris: Pitois-Levraux, 1839. Partial text in M. E. Chevreul, The principles of harmony and contrast of colors, F. Birren, ed., West Chester, PA: Schiffer, 1987. W. Benson, *Principles of the Science of Colours Concisely Stated to Aid and Promote Their Useful Application in the Decorative Arts*, London: Chapman & Hall, 1868. A. Kirschmann, Color-saturation and its quantitative relations, *American Journal of Psychology* 7 (1895) 386–404.
31. J. H. Lambert, *Photometria, sive de mensura et gradibus luminis, colorum et umbra*, Augsburg, Germany, 1760.

32. A. König, *Gesammelte Abhandlungen zur physiologischen Optik*, Leipzig: Barth, 1903.
33. The source of the Schrödinger figure is found in Note 24.
34. An up-to-date general source for information on the neurophysiological organization of color vision is found in K. R. Gegenfurtner and D. C. Kiper, Color vision, *Annual Review of Neuroscience* 26 (2003) 181–206.
35. See note 25 for reference.

Chapter 11

1. For a history of thinking on art and beauty, see A. Hofstadter and R. Kuhns, eds., *Philosophies of Art and Beauty*, Chicago: University of Chicago Press, 1964.
2. V. S. Ramachandran and W. Hirstein, The science of art: a neurological theory of aestetic experience, *Journal of Consciousness Studies* 6 (1999) 15–51.
3. S. Zeki, *Inner Vision*, Oxford: Oxford University Press, 1999.
4. N. Humphrey, Cave art, autism, and the evolution of the human mind, *Journal of Consciousness Studies* 6 (1999) 116–123.
5. *C. Plinii Secundi historiae mundi libri XXXVII*, Basel: Froben, 1539. Pliny the Elder, *Natural History*, Cambridge, MA: Harvard University Press, 1938.
6. For the achievement of Suger, see J. Gage, *Color and Culture*, Boston: Bulfinch Press, 1993.
7. Translations of several of these manuscripts, including that by Eraclius, are found in M. P. Merrrifield, *Medieval and Renaissance Treatises on the Arts of Painting*, Mineola, NY: Dover, 1967.
8. C. Cennini, *The Craftsman's Handbook: "Il libro dell'arte,"* transl. D. V. Thompson, New Haven, CT: Yale University Press, 1933. L. B. Alberti, *Leon Battista Alberti on Painting*, J. Spenser, ed., New Haven: Yale University Press, 1956. Leonardo da Vinci, *The Notebooks of Leonardo da Vinci*, transl. E. MacCurdy, New York: Reynal and Hitchcock, 19 G. P. Lomazzo, *Trattato dell'Arte della Pittura, Scultura ed Architectura*, 3 vols., Milan, 1584. An excellent English edition of Vasari is G. Vasari, *Lives of the Most Eminent Painters, Sculptors, and Architects*, 3 Vols., New York: Abrams, 1973. R. Borghini, *Il Riposo*, Florence, 1584.
9. Original texts of the mentioned works by Dolce, Occolti, and Calli can be viewed at www.bivionline.it.
10. An excellent recent book on Poussin is A. Mérot, *Nicolas Poussin*, New York: Abbeville, 1990.
11. A copy of the painting can be seen in J. Gage, *Color and Culture*, Boston: Bulfinch Press, 1993.
12. Anonymous, *Traité de la Peinture en Mignature*, The Hague: van Dole, 1708.
13. A copy of this painting can be viewed in J. Gage, *Color and Culture*, Boston: Bulfinch Press, 1993.
14. As reported in G. Sand, *Impressions et Souvenirs*, Paris: Lévy, 1873.
15. As reported in H. Delaborde, *Notes et Pensées de J. A. D. Ingres*, Paris, 1984.
16. For a discussion of the optical effects of divisionism, see F. Ratliff, *Paul Signac and Color in Neo-Impressionism*, New York: Rockefeller University Press, 1992.
17. W. Kandinsky, *Über das Geistige in der Kunst*, Bern: Benteli, 1911.

18. P. Klee, *Ueber die moderne Kunst*, Bern: Benteli, 1949.
19. J. Itten, *The Art of Color*, New York: Wiley, 1973. J. Albers, *Interaction of Color*, New Haven, CT: Yale University Press, 1963, also obtainable on interactive compact disk.
20. For a modern description in terms of neurobiology of some of these effects see M. Livingstone, *Vision and Art, The Biology of Seeing*, New York: Abrams, 2002.

Chapter 12

1. R. Fludd, *Philosophia Sacra*, Frankfurt, 1626.
2. S. Kobayashi, *A Book of Colors*, Tokyo: Kodansha, 1987.
3. For literature on synesthesia, see Chapter 2, Note 11.
4. J. L. Hoffmann, *Versuch einer Geschichte der malerischen Harmonie überhaupt und der Farbenharmonie insbesondere*, Halle, Belgium, 1786.
5. W. Preyer, Die Verwandtschaft der Töne und Farben, *Jenaische Zeitschrift für Medizin und Naturwissenschaften* (1870) 759–765.
6. F. W. Unger, Über die Theorie der Farbenharmonie, *Poggendorffs Annalen der Physik und Chemie* 87 (1852) 121–128.
7. V. Goldschmidt, *Farben in der Kunst*, Heidelberg: Winter, 1919.
8. B. Thompson, Conjectures respecting the principles of the harmony of colours. *In Philosophical Papers*, Vol. 1, London: Cadell and Davies, 1802.
9. As described in T. M. Cleland, *A Grammar of Color*, Mitteneague, MA: Strathmore Paper Company, 1921.
10. W. Kandinsky, *Über das Geistige in der Kunst*, Bern: Benteli, 1911.
11. J. Itten, *The Art of Color*, New York: Wiley, 1973.
12. This chapter has profited from the study of A. Schwarz, *Die Lehren von der Farbenharmonie*, Göttingen, Germany: Muster-Schmidt, 1999.

Glossary

absorbance a measure of the absorption of light by a colorant; reciprocal of transmittance or reflectance.

absorption the transfer of energy from photons to matter.

achromatic neutral, possessing no hue and chroma.

adaptation adjustment of response to wide ranges of stimulus. There are chromatic and brightness/lightness adaptation.

afterimage a visual image experienced after its original stimulus has ended.

aim color a psychophysical color specification for a color chip; typically in a systematic collection.

amacrine cells a cell type in the retina between the bipolar and the ganglion cell layer.

angle of incidence angle at which light rays impinge on a surface, expressed with respect to a line orthogonal to the surface.

angle of viewing the polar angle describing the difference between the lines connecting the retina with the outer limits of an observed patch, expressed in degrees. It is a function of patch size and distance from the eye.

array detector system a series of light-sensitive devices arranged in a way that each detector is exposed to light of a narrow range of wavelengths. Used in some types of color-measuring equipment.

attention focusing process of the brain on a small portion of information acquired by the senses.

Color: *An Introduction to Practice and Principles, Second Edition*, by Rolf G. Kuehni
ISBN 0471-66006-X Copyright © 2005 John Wiley & Sons, Inc.

auxochrome color enhancer, a chemical group attached to a chromophore molecule to produce enhanced light absorption in the visible range.

Bezold–Brücke effect a sensory effect named after two German scientists, according to which hue sensations caused by light of all but three wavelengths change with changing light intensity.

bipolar cells a cell type in the retina that receives input from the cones and creates output to ganglion cells.

blackbody an idealized, nonreal material that absorbs and emits energy at all wavelengths without restriction.

brightness attribute of visual sensation, according to which a given visual stimulus appears more or less intense. Differences in brightness range from dim to bright.

cangiante a changeable optical effect when viewing fabrics woven with warp and weft yarns of contrasting colors, resulting in simultaneous multiple colors, depending on the angle of the fabric.

charge transfer movement of an electron usually attached to one atom over to another. The movement represents an excited state, but the movement satisfies the overall electrical balance of crystalline arrangements of certain atoms.

chemical bond the attraction that keeps atoms locked together in a molecule.

chroma attribute of sensation permitting the judgment of the degree to which a chromatic object color differs from the achromatic color of the same lightness. A measure of chromatic intensity.

chromaticity diagram a two-dimensional psychophysical diagram representing the chromatic component of a color stimulus.

chromaticness attribute of visual sensation, according to which a patch appears to exhibit more or less chromatic color.

chromophore color carrier, an organic molecule absorbing specific wavelength bands of visible light.

CIE colorimetric system a color-stimulus-specification system developed by the International Commission on Illumination (CIE is the acronym of the French name of the organization, Commission Internationale de l'Éclairage).

cleavage the quality of a crystallized substance of splitting along specific internal planes.

colorant a material that changes the absorption characteristics of another material: dyes, pigments, or dissolved metal salts.

colorant trace the trace in a geometrical color space, such as the CIE x, y, Y space, resulting from the connection with a line of individual stimulus loci in that space, representing application of the colorant in multiple concentrations.

color attributes fundamental aspects of color perception, for example, hue, chroma, lightness, or hue, whiteness, blackness.

color constancy lack of change in the apparent color of an object regardless of quality or quantity of the illuminating light. Natural objects tend to be reasonably constant in appearance when viewed after adaptation to various phases of daylight.

color difference perceived difference between two nonidentical fields of color.

color difference formula a psychophysical mathematical formula that allows the calculation of the approximate average perceived difference between two stimuli.

color fidelity perceived degree to which a color in a reproduction matches the original color.

color harmony the combination of color elements in a work of art or craft so that the total effect is perceived as being in concord.

colorimetry the branch of color science concerned with the psychophysical numerical specification of color stimuli.

color-matching functions three spectral functions describing the number of three real or nonreal lights required to cause color perceptions matching those of spectral lights.

color metric a metric describes the mathematical internal structure of a geometrical space; a color metric applies to a psychophysical color space.

color, primary colloquial term for one of three lights, the appearance of which cannot be matched by mixing the other two, and that is used in combination with the other two to produce stimuli resulting in matching sensations for any visible light stimulus. The term is also colloquially used to designate one of three colorants used in color reproduction systems (yellow, red, and blue, or yellow, magenta, and cyan).

color, related color perception caused by light reflected from objects in the presence of other objects. The perceived color is dependent on the colors of the surrounding objects.

color rendering the fidelity with which artificial light sources render the appearance of colored objects in comparison with a standard daylight.

color scale a scale in which perceived colors change in a systematic manner, usually in one attribute.

color space three-dimensional geometrical construct with a coordinate system based directly or indirectly on average cone sensitivity functions and housing all possible color perceptions in a systematic manner.

color stimulus a stimulus is something that excites an organism, or one of its components, to functional activity; the external color stimulus normally consists of light of one or more wavelengths.

color, unrelated a perceived color that fills the complete field of view of the eye, or that is isolated, such as a colored material viewed through a reduction screen (black screen with a small opening).

cone response functions a set of three spectral functions or series of numbers describing the absolute or relative response of cones to light of different wavelengths.

cone, visual a type of cone-shaped light-sensitive cell in the retina. In trichromatic vision there are three cone types differing in their spectral sensitivity.

contrast, simultaneous a change in apparent lightness, hue, and/or chroma of a colored field caused by an adjacent or surrounding field of different lightness, hue, and/or chroma. Both fields change appearance in a direction away from the color of the other field; for example, a light-gray field surrounded by a deep red field looks greenish and lighter than when viewed in a white surround.

contrast, successive an imaginary colored field perceived in a location in the visual space where previously a colored object was located and viewed attentively for a time. The afterimage is either negative (say, red after viewing a green object) or positive. Afterimages can be experienced with open or closed eyes.

corpuscular theory of light one of two accepted theories of light according to which light consists of unit "packages" of electromagnetic energy, called quanta or photons.

correlated color temperature absolute temperature on the Kelvin scale of a blackbody emitting light that gives rise to the same color perception as a given test light. Used to describe an aspect of the quality of lamp-light with spectral power distributions different from those of an emitting blackbody.

crispening effect contrast effect involving brightness/lightness, as well as chromaticity. The smallest change in stimuli to obtain a just noticeable difference between two color fields, if their lightness and chromaticity straddle those of the surrounding area.

diffraction the modulation of a wave passing the edge of an opaque material, resulting in a redistribution of energy due to bending of waves.

dyes natural or artificial colorants that absorb, but usually do not scatter light and are soluble in the substrate or that go through a solution stage in their application to the substrate.

electromagnetic radiation transport of electromagnetic energy through space; electromagnetic radiation has a wide spectrum, from X rays to radio and television transmission rays.

electron volt unit of energy equal to the energy acquired by an electron falling through a difference in electric potential of one volt.

emission–immission theory an early theory of vision, promoted by the Greek philosopher Plato, according to which energy flows from the eye to the objects (emission) and from the objects to the eye (immission), resulting in vision.

Euclidean space space in which Euclid's axioms of straight and parallel lines and angles of plane figures apply, for example, a cube.

fluorescence a form of luminescence in which ultraviolet or visible light is absorbed and emitted at a higher but visible wavelength. Fluorescence ceases once the flow of arriving photons stops.

fovea a small depression in the retina containing mainly cones at high density. It is the most sensitive area of the retina, that on which the optical image at the center of our gaze is focused.

gamut a region of color space occupied by stimuli resulting from colorations of two or more colorants, for example, three dyes in color photography.

ganglion cell a cell type in the retina whose output fibers form the optic nerve.

halftone printing a printing process where the image is rendered in smooth variations of dots of the same or varying size of the four (or more) process inks yellow, magenta, cyan, and black.

hue attribute of color perception denoted by the names yellow, red, blue, green, and so forth.

hue superimportance relates to the fact that in a Euclidean color space contours of unit perceptual difference from a central reference are always elongated and pointing in the direction of the achromatic axis of the space. The implication is that a smaller change in stimulus is required for a hue difference than for a chroma difference of equal perceived magnitude (factor approximately 1:2).

hue, unique a hue that cannot be described by hue names other than its own. There are four unique hues that have no perceptual similarity to any of the other three: yellow, red, blue, and green. A color of unique red hue has neither a yellowish nor a bluish appearance, and comparably for the other three.

illuminant the numerical qualitative description of a light source in the form of its spectral power distribution.

illuminant, equal energy an illuminant having the uniform spectral power distribution of 1 across the spectrum.

incandescence the emission of visible light from a body at high temperature (above ca. 1500 K).

index of metamerism a numerical index representing a measure of the magnitude of the degree of mismatch of two metameric objects viewed in a test light that is different from the reference light.

integrating sphere a hollow spherical device on color-measuring instruments, coated on the inside with white, highly reflecting material uniformly dispersing the light of a lamp onto a sample to be measured for reflectance properties, making possible the uniform sampling of average scattered light.

interference the process in which two or more electromagnetic waves combine or cancel each other, depending on if they are in or out of phase when combined.

intra-, inter- prefixes, the former with the meaning of within, the latter of among.

isotropic exhibiting properties with the same values when measured along axes in all directions; in connection with color space: a space in which distances in all directions are commensurate with the size of perceived distances; generally: a uniform color space.

iterative technique a mathematical method in which results of calculations are improved stepwise by taking the previous solution as a starting point for finding an improved solution. The steps are continued until no further improvement is obtained or until the quality of the solution meets requirements.

Kelvin scale temperature scale with Celsius scale units, but beginning at absolute zero (−273.2°C).

Kubelka–Munk relationship relates the reflectance of a partly absorbing and partly scattering object to absorption and scattering, for example, for a layer of paint or a dyed textile material.

lateral geniculate nucleus a mass of cells in the brain along the visual passageway between the eye and visual areas at the back of the brain. There are two such nuclei, and significant processing of signals is known to take place there.

lightness attribute of a visual sensation involving object colors, according to which a color field appears to emit equal or less light compared to a perfectly white field. Lightness can be understood as relative brightness. Differences in lightness range from dark to light.

luminance photometric quantity measure of light reflecting the average human sensitivity to light. Its unit is candela per square meter.

luminescence the emission of light from a body below the temperature of incandescence: chemiluminescence is caused by certain chemical reactions; photoluminescence refers to absorption of light at a lower wavelength and emission at a higher but visible wavelength (see also fluorescence and phosphorescence).

luminous reflectance the product of the reflectance of an object, spectral power of a light source, and the luminosity function (spectral daylight sensitivity) of the standard observer; represented by the CIE tristimulus value Y.

magnocellular system component of the visual system of primates responsible for perception of motion. Its cells are comparatively large, thus its name.

masstone the perceived color (or its stimulus) of a pure pigment dispersed in a paint medium and applied in an opaque layer onto a substrate.

matching manipulation of color stimuli or colorants so that the combined stimulus or colorants result in a perception identical to that of a standard stimulus or coloration.

mathematical transformation to change the form or direction of a point, line, area, or solid from one set of coordinates to another set without changing the intrinsic information. Based on Grassmann's laws, color-matching functions can be transformed linearly to change their shapes without loosing the intrinsic information contained in them.

metamerism the property of spectral power distributions to have the same impact on the visual system despite the fact that the spectral distributions are different. Metameric lights or objects have identical tristimulus values despite differences in spectral power distributions or reflectance functions.

monochromatic, polychromatic a visible light stimulus consisting of a single wavelength or a very narrow range of wavelengths is called monochromatic; it is called polychromatic if it has a spectral power distribution broader than a narrow band.

monochromator a physical apparatus allowing the separation of polychromatic light into its monochromatic components, for example, a prism or a grating.

mordant a chemical applied to fiber material interacting chemically with certain dyes (depending on their chemical class) so that they become fixed to the fiber.

nanometer metric unit of distance measurement, a nanometer is a billionth of a meter. There are approximately 25 million nanometers in an inch.

neural network refers to a structurally integrated unit of the brain, as well as to a mathematical procedure imitating its presumed activity on a digital computer. Neural networks learn to extract fundamental information from input data and use it to generate outputs at a cognitively higher level.

neurophysiology the physiology (science of the organic processes of living systems) of the nervous system.

opaque property of a material through which light cannot pass; not transparent or translucent.

opponent color theory the theory according to which signals from the three types of cone cells are subtracted in pairs to form opponent signals (either positive or negative). Cells with opponent character have been found in the retina and the lateral geniculate nuclei.

parvocellular system component of the visual system of primates responsible for the detection of structural details and color. Its cells are comparatively small.

partitive color mixture a type of additive color mixture where the mixing results from rapid superimposition of different stimuli, such as sectors of differently colored papers on a rapidly rotating disk (disk mixture) or from spatial integration of multiple and different small stimuli (color television, halftone printing).

perception subjective, conscious experience of the impact of an outside force on a sensory system.

phosphorescence a form of luminescence in which ultraviolet or visible energy is absorbed by a material and emitted at a higher but visible wavelength. Phosphorescence persists for a certain length of time after the flow of absorbed photons stops.

photometer instrument for the measurement of light intensity.

photon unit or quantum of electromagnetic radiation.

photopic vision vision mediated by cones active at daylight levels of light, and resulting in brightness and color perceptions

pigments natural or artificial colorants that not only absorb but also scatter light and that are insoluble in the application medium or substrate.

pointillism a style of painting developed by the French painter Georges Seurat where small dots or dashes of different colors are placed side by side with the (erroneus) expectation that, when viewed at a distance, they will additively fuse, resulting in brighter colors.

polychrome used to designate a work of art or craft containing more than one color stimulus.

psychophysics the branch of psychology that attempts to elucidate the relationship between stimulus and a resulting perception.

qualia (plural of quale) presumed qualitative fundamentals of brain/mind activity (comparable to quanta being quantitative fundamentals of electromagnetic energy). Qualia are, for example, the redness of red, the sweetness of the taste of sugar. There is ongoing discussion concerning the existence of qualia.

quantum (plural: quanta) the smallest increment of electromagnetic radiation (see also *photon*).

reflectance factor is the ratio of light of a specific wavelength reflected under certain conditions from a given surface to that same light reflected from a perfectly diffusing surface; it is the result of reflectance measurement.

reflection the process of returning electromagnetic radiation from a reflecting surface. The radiation is returned according to simple optical laws (the angle of reflection from the surface equals the angle of incidence on the surface). Reflection is the optical principle on which the reflecting or mirror telescope is based.

refraction change in direction of a beam of electromagnetic radiation due to change from one medium (say, air) into another (say, glass) in which its speed of propagation differs. Refraction is the optical principle on which the refracting or lens telescope is based.

retina a layer of cells coating the inside rear wall of the camera-type eye, containing light-sensitive rod and cone cells, as well as several cell types connected to rods and cones. The retina is continuous with the optical nerve, and carries information to the visual center in the brain.

retinal a natural dye attached to one of four similar but different protein molecules to form the light-sensitive mechanisms in rods and cones in the primate visual system. In its active form, it has a purple color. Under the influence of light it changes its configuration and becomes colorless and temporarily inactive.

rods rod-shaped type of light-sensitive cells in the retina responsible for night (scotopic) vision.

saturation attribute of visual perception indicating the degree to which a chromatic sensation differs from an achromatic sensation, regardless of its perceived brightness.

scattering the process of returning electromagnetic radiation by a scattering surface. Scattering surfaces return light in all directions, resulting in diffusion of the light beam.

scotopic vision vision without color mediated by rods in the retina; highly sensitive and responsible for night vision capabilities.

sensation, perception in classic psychology the former term referred to immediate and direct qualitative experiences, while perception referred to additional psychological processes involving memory, meaning, and judgment. This differentiation has now been largely abandoned and in this text the term perception is mainly used.

spectral power distribution the relative amount of light at a different wavelength of a polychromatic light with (usually) light at 555 nm = 1.0.

spectral sensitivity sensitivity of a light-sensitive detector at different wavelengths.

spectral space space representing ordering of color spectra (such as the reflectance spectra of Munsell color chips) by the method of dimension reduction. Unlike psychophysical color spaces, they are not directly related to human color vision.

spectrophotometer instrument for measuring the spectral reflectance of objects.

spectroradiometer instrument for the determination of the spectral intensity of lights.

spectrum, visible array of lights of visible wavelengths arranged according to wavelength.

standard error statistical term for a measure of the average deviation from the mean value in a series of values that form a bell-shaped curve when plotted; if the bell shape is wider, the standard error is larger, and vice versa.

synaesthesia concomitant sensation; there are several different cases of synaesthesia, for example, color sensations accompanying perceived sounds, or vice versa.

tetrachromacy color vision system with four different types of cones. Approximately half of human females have the genetic potential for tetrachromacy. It is also found in some other animal species.

tint-shade scale a scale of colors beginning at white, passing through mixtures with increasing amounts of a chromatic pigment to the full color (color of highest chroma), and passing through mixtures with increasing amounts of black to black. A tint-shade scale is of constant hue but varying in lightness and chroma.

tonal scale for the purposes of this text, tonal scale is defined as a constant hue, constant lightness scale varying in chroma.

transmission passage of light through a transparent medium, for example, through glass or a liquid.

translucent attribute of a material that is neither transparent nor opaque, such as a frosted glass panel.

transparent attribute of a material that allows light of some or all wavelengths to pass through it.

trichromacy color vision system with three different cone types found in the majority of primates.

tristimulus values amounts of three primary lights superimposed to match the perception created by any light. In the CIE colorimetric system, trichromatic values are named X, Y, and Z and refer to imaginary lights. Tristimulus values are used for specification of color stimuli.

uniform chromaticity diagram two-dimensional diagram in which geometrical distances between points representing color perceptions correspond to their perceptual distances.

uniform color space a three-dimensional geometrical space in which geometrical distances between points representing color perceptions correspond to their perceptual distances (see also isotropic).

value attribute designation in the Munsell color system corresponding to lightness.

vector a quantity with magnitude and direction usually represented by an arrow, the length of which represents magnitude and its position in space indicates the direction of the magnitude.

wavelength distance in the direction of propagation between two peaks of a wave.

xerography image reproduction process in which the image is represented by static electricity on a sheet of paper attracting resinous pigments that then are permanently fused to the paper.

Credits

Permission to reproduce certain figures was obtained from the copyright holders and is gratefully acknowledged:

Elsevier Science Inc., New York, NY: Figure 3.4 reprinted from *Vision Research* 40 (2000), R. A. Rensink, Seeing, sensing and scrutinizing, p. 1485.

Illuminating Engineering Society, New York, NY: Figure 1.1 reprinted from, *IES Lighting Handbook*, 1972.

Optical Society of America, Washington, DC:

Figure 3.8 reprinted from *Journal of the Optical Society of America* 45 (1955), D. Jameson and L. M. Hurvich, some quantitative aspects of an opponent-colors theory, p. 549.

Figure 5.14 reprinted from *Journal of the Optical Society of America A* 13 (1996), R. Lenz, M. Osterberg, J. Hiltunen, T. Jaaskelainen, J. Parkkinen, unsupervised filtering of color spectra, p. 1317.

Figure 7.6 reprinted from *Journal of the Optical Society of America* 64 (1974), D. L. MacAdam, Uniform color scales, p. 1696.

Figure 5.10 reprinted from *Journal of the Optical Society of America* 68 (1978), C. E. Foss, Space lattice used to sample color space, p. 1616.

Sinauer Associates Inc., Sunderland, MA: Figure 4.1 reprinted from D. Purves and R. B. Lotto, *Why We See What We Do*, 2003, p. 57.

Springer-Verlag, Heidelberg, Germany:

Figure 3.5 reprinted from D. Jameson, *Handbook of Sensory Physiology*, Vol. 7/4, 1972.

Figures 8.2 and 8.3 reprinted from D. L. MacAdam, *Color Measurement*, 1981, pp. 112 and 113.

Color: *An Introduction to Practice and Principles, Second Edition*, by Rolf G. Kuehni
ISBN 0471-66006-X Copyright © 2005 John Wiley & Sons, Inc.

John Wiley & Sons, Inc., Hoboken, NJ:

Figures 5. 7a and 5.7b reprinted from *Color Research and Application* 21 (1996), A. Hård, L. Sivik, and G. Tonnquist, NCS-Natural Color System p. 180–205.

Figures 5.8b, 6.6, 6.9, 6.11, 7.7, 7.8 reprinted from G. Wyszecki and W. S. Stiles, *Color Science*, 2nd ed., 1982, pp. 137, 184, 510.

Figure 5.13 reprinted from *Color Research and Application* 6 (1981), D. Nickerson, OSA Uniform Color Samples: a unique set, pp.1–33.

Figures 7.10 and 7.11 reprinted from *Color Research and Application* 26 (2001) M. R. Luo, G. Cui, and B. Rigg, The development of the CIE colour-difference formula CIEDE2000, p. 346.

Figure 8.7 reprinted from R. M. Johnston, Color theory, in T. A. Lewis, ed., *Pigments Handbook*, Vol. 3.

Figure 8.8 reprinted from F. W. Billmeyer and M. Saltzman, *Principles of Color Technology*, 2nd ed., 1981, p. 158.

Index